AN ECOLOGICAL AND CULTURAL CRITIQUE OF THE COMMON CORE CURRICULUM

Studies in the Postmodern Theory of Education

Shirley R. Steinberg
General Editor

Vol. 471

The Counterpoints series is part of the Peter Lang Education list.
Every volume is peer reviewed and meets
the highest quality standards for content and production.

PETER LANG
New York • Bern • Frankfurt • Berlin
Brussels • Vienna • Oxford • Warsaw

CHET BOWERS

AN ECOLOGICAL AND CULTURAL CRITIQUE OF THE COMMON CORE CURRICULUM

PETER LANG
New York • Bern • Frankfurt • Berlin
Brussels • Vienna • Oxford • Warsaw

Library of Congress Cataloging-in-Publication Data

Bowers, C. A.
An ecological and cultural critique of the common core curriculum / by Chet Bowers.
pages cm. — (Counterpoints: studies in the postmodern theory of education; v. 471)
Includes bibliographical references.
1. Educational change—Social aspects—United States.
2. Education—Standards—Social aspects—United States.
3. Education—Curricula—Social aspects—United States.
4. Environmental education. I. Bowers, C. A. II. Title.
LC191.4.B68 379.1'58—dc23 2014031263
ISBN 978-1-4331-2799-1 (hardcover)
ISBN 978-1-4331-2798-4 (paperback)
ISBN 978-1-4539-1402-1 (e-book)
ISSN 1058-1634

Bibliographic information published by **Die Deutsche Nationalbibliothek.**
Die Deutsche Nationalbibliothek lists this publication in the "Deutsche Nationalbibliografie"; detailed bibliographic data are available on the Internet at http://dnb.d-nb.de/.

The paper in this book meets the guidelines for permanence and durability of the Committee on Production Guidelines for Book Longevity of the Council of Library Resources.

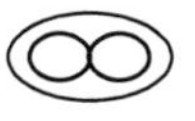

29 Broadway, 18th floor, New York, NY 10006
www.peterlang.com

Printed in the United States of America

Contents

Preface

The Common Core reforms are being heralded by heads of corporations, the computer industry, and governmental officials as necessary to making America more competitive in the global economy. I shall be focusing on why these corporate- and government-sponsored reforms (Common Core Standards, hereafter referred to as CCS) are the wrong ones in this era of overshooting the sustaining capacity of natural systems. This is not understood by the promoters of the CCS reforms that have been adopted by the majority of states, and by politicians in states such as Texas and Oklahoma who want to establish their own state standards. Nor is it understood by powerful, so-called conservative critics.

The primary criticism of past approaches to education in America—whether it is an approach based on the ideas of John Dewey and other progressive theorists, the liberal arts and Great Books approach advocated by Mortimer Adler and Robert Hutchins, or the alternative approaches of Montessori and charter schools—is that they all have ignored how the language used in the different approaches to education marginalize awareness of environmental limits. Greater reliance upon a liberal education (informed by the insights of the great thinkers from the past), the students' supposed self-directed learning and own intuitive wisdom in constructing their own knowledge, and the various approaches to free classrooms and biblical-based learning have not addressed the cultural roots of the ecological crisis.

In effect, the CCS reforms are a continuation of America's long history of miseducation, which has resulted in a large number of Americans denying that there is an ecological crisis. This state of denial, which seems unaffected by the evidence of extreme weather patterns, droughts, massive forest fires, and the increasing number of environmental refugees trying to move across national borders, is leading Republican politicians at both the state and federal levels to urge the defunding of research on how humans are contributing to climate change. This ideologically driven denial will be strengthened by the CCS reforms and by the growing trend of major universities offering online degrees—which leads to another set of problems that accompany computer-mediated learning in an increasingly surveillance-driven culture.

In effect, both the continued miseducation that leaves unchallenged the ecologically problematic cultural assumptions as well as the fundamental changes in the basic foundations of daily life introduced by the digital revolution are turning democracy into a relic of the past, as democracy requires informed decision making in face-to-face communities. The areas of social life where democracy is undergoing a revitalization are among the grassroots groups working to reduce their reliance upon processed foods, the addiction to consumerism, the mind-altering advertising industry, and to being less seduced into equating progress and happiness with amassing material wealth that requires destroying the self-renewing capacity of natural systems.

The recognition of environmental limits is strengthening the awareness that people must work together in understanding the limits and possibilities of their local environments, and the importance of reaching agreements on how to live more sustainable and community-centered lives. The majority of Americans, however, have been educated to take for granted the cultural assumptions that gave conceptual direction and moral legitimacy to the technological developments and market forces that have led to our consumer-dependent lifestyles. Unfortunately, the market liberal and libertarian ideologues in denial that there is an ecological crisis, and that human behavior may be a major contributor, have been given by the Supreme Court the go-ahead to make unlimited donations in the hundreds of millions of dollars to ensure the election of politicians who will promote their corporate agendas.

The two main foci of my critique of the CCS reforms are that they will further impede awareness of the cultural roots of the ecological crisis, and that they will undermine the possibility that students will acquire the conceptual basis necessary for recognizing how the ecologically problematic assumptions inherited from the past continue to be reinforced. Also undermined will be the students' understanding of ecologically sustainable community-centered patterns that need to be revitalized. If students lack these understandings, they will be unable to exercise communicative competence in the most critical areas of social life—which in the years immediately ahead will be marked by increasing unemployment as more work becomes computerized and as more people slip below the poverty line.

Before discussing both the misconceptions underlying CCS reforms and what students should be learning if they are to have a voice in making the transition to ecologically sustainable communities, it is necessary to consider the many radical and life-altering changes introduced by computer scientists and corporations that have bypassed the democratic process. The majority of Americans, like a Greek chorus, have cheered these changes, except when their identities have been stolen and their jobs have been taken over by robots and computer-driven systems. Besides, who needs democracy when the virtual realities made possible by digital technologies create freedoms unimagined in the past?

There is a powerful group of computer scientists who have morphed into futurist thinkers working to create a radically different form of progress that may be a surprise to most people, unless they have already read Ray Kurzweil's book *The*

Age of Spiritual Machines (1999). Kurzweil and his colleagues assume that we are entering a new epic in the evolution of human life, where super-intelligent computers will shortly displace Nature's need for humans. Kurzweil and his colleagues are working to develop the technologies that will make this transformation possible. Like so many other radical changes being introduced by the digital revolution, their efforts to be both the oracle and implementers of Darwin's theory (as they interpret it) are not subject to the democratic process.

On a less epic level of radical and irreversible changes that have already bypassed the democratic process are the widespread technologies introduced by the digital revolution that people have willingly embraced. Personal convenience, efficiency in performing tasks, quick access to information, and a thousand other uses (many of them positive) have led to a state of passive awareness that has left the most important questions unasked. These include how far we should go in computerizing the workplace, in undermining privacy, in allowing massive amounts of data on people's behavior to be collected and sold by data brokers, in ignoring the increased vulnerability of people's savings and the threats to the country's critical infrastructures, which can be brought down by skilled and determined hackers—or in cyber attacks from foreign soil. In short, the irreversible changes brought about by the computer industry and corporations have not increased our well-being and safety as individuals and communities. Yet, the public continues to exhibit a split personality, both equating these changes with progress and being passive about the increased dangers to their lives.

I will devote considerable attention to what teachers, heads of corporations (including Bill Gates, who envisions huge profits for Microsoft in providing the software upon which the CCS reforms depend), and both liberal and so-called conservative politicians should have learned during their years in universities. Of more immediate concern is whether the so-called conservative critics of the CCS reforms have also been victims of a colossal process of miseducation that has its roots in what they regard as the near-sacred writings of the men who gave us the 100 Great Books and other literary works they identify as the highest achievements of the human mind. What is especially notable about the 100 Great Books, which often form the primary basis of a liberal education, is that none of them mention that there are environmental limits, that their abstract patterns of thinking undermine the exercise of ecological intelligence and the dangers of hubris that lead to ignoring other cultural ways of knowing—including what could be learned from other cultures about living within the limits and possibilities of their bioregions.

The Critique of CCS Reforms by So-Called Conservatives

Perhaps the most articulate and thus most important self-labeled conservative critic of the CCS reforms is Terrence O. Moore, a professor of history at Hillsdale College in Michigan and the author of *Story Killers: A Common-Sense Case Against the Common Core* (2013). He is also a chief promoter of the classical charter school movement. Many of his criticisms of both the CCS and the progressive reforms that preceded it over the past generations are valid, including the criticism that teachers project onto students their own lowered intellectual expectations and fail to introduce students to the range of cultural achievements in literature, the arts, and politics that would have provided the basis of a more informed and personally enriched life. He also is correct in his observation that the educational background of too many teachers leaves them unable to recognize the difference between good literature and what is simply easy to read. This criticism, while valid in so many ways, fails to take account of the poverty, homelessness, and other factors that lead to classrooms of students unprepared to make the transition from their largely oral cultures to print culture and thus the abstract thinking required in most classrooms. The stereotyping of students and the practice of warehousing students deemed less able are also part of the problem that Moore does not face at Hillsdale College or in the classical charter schools he promotes.

However, the more important issue is whether the intellectually rigorous mind- and character-building encounters with the highest achievements in Western culture that he argues are the conservative's alternative to the corporate-sponsored CCS reforms will continue the process of miseducation. This is an especially important question, as we have now entered an environmentally and population-stressed era where the consequences of previous silences and missteps in a public school and university education cannot so easily be ignored or compensated for with massive governmental programs.

My reference to "so-called conservatives" needs to be explained as it is fundamental to the larger question of what Moore wants the educational reforms to conserve. It is also important to rectifying our political vocabulary, as we are now operating in the political world George Orwell described in his dystopian classic *Nineteen Eighty-Four*. Identifying what Moore wants to conserve requires recognizing his status as a faculty member of Hillsdale College, as well as the importance of Hillsdale College as a seedbed of ideas that support the "original intent" interpretation of the framers of the Constitution (which supports corporations by limiting the role of government in addressing a wide range of abuses). Another example of Moore's faux conservatism, and thus the faux conservatism of Hillsdale College, is their support of corporate capitalism, and their silence on the importance of social justice issues. And most important of all, we need to consider the college's ideologically driven silences on what needs to be conserved as the ecological crisis deepens and as the digital revolution takes us further down the pathway leading to a police state.

Market liberalism is the label that best describes the ideology of Moore and other so-called conservatives, such as the five Supreme Court judges who voted that unlimited spending on political campaigns is a form of free speech. The word *conservative* should be used to describe the land conservancy movement, and the people who defend civil liberties from encroachment by the corporations and computer scientists who are attempting to elevate data as the highest form of knowledge, which now leads to monetizing the data collected on people's lives. Other groups deserving the label *conservative* include the people carrying forward the intergenerational knowledge and skills (that is the cultural commons) that represent alternatives to the consumer-dependent existence promoted by corporations in denial about the ecological impact of their carbon and toxic footprints, and those working to conserve and extend the social justice achievements of the past. These conservatives draw their inspiration not from the writings of Ayn Rand and Milton Friedman but instead from Edmund Burke, Michael Oakeshott, Mahatma Gandhi,Vandana Shiva, and Wendell Berry.

One more point, which relates more directly to Moore's argument that the educational reform now needed by the nation is the liberal education that has nurtured the creativity of artists, writers, and leaders in business and government, and inspired others with an understanding of America's greatness: As I recall from reading Western philosophers and political theorists, none were aware of environmental limits, or of their own ethnocentrism. Nor did their hubris lead to recognizing how their abstract theories marginalized awareness of the intergenerational, face-to-face achievements across a wide range of human activities that enable people to live less money-dependent lives. My own experience of having observed over the last 50 years the leading roles played in government, major corporations, and other powerful institutions by graduates of elite universities strongly suggests that the liberal arts curriculum advocated by Moore does not strengthen democracy and an awareness of the dangers of colonizing other cultures. The policies of John Foster Dulles and Allen Dulles, both graduates of an elite university and a liberal education, are atypical only in the power they were able to exercise in undermining local democracies and colonizing other cultures. It is also necessary to consider whether the liberal education students encountered at elite universities such as Princeton and Yale led them to seek careers on Wall Street or in the CIA, as heads of major corporations that outsource both their production processes and corporate earnings in order to avoid paying taxes. Most approaches to a liberal education have ignored examining the deep cultural assumptions taken for granted by Western philosophers and social theorists. Few graduates who have received the liberal education advocated by Moore understood how the language inherited from these abstract thinkers continues to reproduce the silences and misconceptions that now guide corporations and universities in achieving their global agendas.

And I suspect that many of the Republicans in Congress as well as the governors supporting the CCS reforms received a liberal education at an elite university, and are now using what they learned to limit the ability of working-class and poor people to vote, to reduce the power of labor unions in an era where robots are displacing the need for workers, and to reduce benefits that were part of earlier social contracts between workers and employers. How many of these so-called conservatives have encountered the conservative ideas of Edmund Burke, Mahatma Gandhi, and Wendell Berry?

Moore's criticisms that CCS reforms are lowering intellectual standards and are excessively focused on transforming students into effective workers who will promote further economic growth need to be taken seriously. The lowered standards of the CCS reforms will result, as he asserts, in students not encountering the best in Western and American literature and creative arts. He wants students to encounter the ideas of Plato, Aristotle, and the other culture-shaping philosophers, well as American classics that address what it means to be human in all its dimensions. But the canon of life-enhancing literature Moore proposes is unlikely to include important African American poets such as Amiri Baraka (formerly Leroi Jones) and Latino poets such as Luis Alberto Urrea. Nor are the writings of Aldo Leopold, Rachel Carson, and Gregory Bateson part of Moore's list of authors who provide models of courage, empathy, and wisdom about the human condition. Moore is correct in his criticism that the CCS reforms will leave students unable to engage in a serious discussion of any important piece of literature. However, his faux conservatism, as well as that of the others who share his ethnocentric biases and silence about the ecological crisis, will ensure that if their educational reforms were implemented, few students would be able to engage in a discussion of the cultural roots of the ecological crisis, including how to bring about the cultural/linguistic changes that will lead people to exercise ecological intelligence. Nor would they understand that the exercise of ecological intelligence is essential to conserving the community and life-sustaining traditions being threatened by market liberal ideologues promoting the digital revolution and the globalization of an individualistic/consumer-dependent lifestyle.

In order to avoid the mistakes of so many critics of the CCS reforms, which ignore how both environmental and technological changes are creating new challenges that cannot be addressed by relying upon the old patterns of thinking, my criticisms will addresses two major problems. The first concerns the misconceptions that teachers are most likely to reproduce in how they teach to the standards that are to be machine-tested (which involves yet another set of problems). The other problem, or what I refer to as the silences in the thinking behind the CCS reforms and in the professional training of teachers, requires an extended discussion of the need to introduce students to basic culturally and ecologically informed understandings that will be essential in the years ahead. Chapter 2, for example, will explain how to introduce students to recognize a basic misconception that is widespread throughout the dominant culture: namely, that there are no ideas, things, events, and even individuals that are autonomous entities. The alternative to this misconception (which is a result of abstract thinking) is to recognize that everything exists in relationships with other participants and entities, and that these relationships serve as information pathways that need to be given attention. This understanding requires being aware of the emergent nature of what occurs in experience.

This, in turn, leads to asking questions about how language frames awareness—including what is not recognized, and how language carries forward the earlier patterns of thinking formed before there was an awareness of environmental limits. This focus on the relational nature of existence in both natural and cultural ecologies also leads to learning to recognize the intergenerational knowledge and skills that exist in every community that relies less on consumerism and that have a smaller adverse ecological footprint. Key concepts are introduced, along with explanations for introducing these concepts when there is space between the required curricula dictated by the CCS.

Educating students to meet the real challenges of the 21st century involves providing them with the language that leads to new insights, and an explicit awareness of cultural patterns and relationships previously not recognized. If certain concepts are continually introduced throughout whatever area of the curriculum students are engaged in, they will begin to recognize the alternatives to the formulaic thinking that threatens their future. I provide many examples of how to introduce students to basic insights, such as recognizing that words have a history and that their current meanings were framed by the analogs settled upon in earlier eras, that print and nouns limit awareness of the relational and emergent nature of life processes occurring in local contexts, that there are nonmonetized activities and relationships that lead to discovering and developing personal talents, that the deep cultural assumptions handed down from the past are largely taken for granted and continually reinforced by others who are unaware of environmental limits, and so forth.

As many of these surprisingly simple yet consciousness-altering insights will be new to most teachers, an effort is made to explain them as fully as possible, and then to devote a chapter to how they can be introduced in the classroom without turning the concepts into yet another unit of the curriculum. The concepts lead to new ways of understanding that are relevant to most areas of the curriculum, particularly since the curriculum is largely based on the deep cultural assumptions taken for granted since the rise of the industrial/consumer-dependent culture.

As the primary focus of the recommendations for reforms that go beyond the test-driven CCS curriculum (as well as the curriculum where the teacher's discretion supposedly plays a larger role) is on enabling students to recognize the relational nature of all aspects of life, ranging from the cultural to the natural environment, there is a chapter that provides an introduction to the wisdom traditions of indigenous cultures as well as those of the world's major religions. Again, the intent is not to introduce students to curriculum units on different wisdom traditions, as few teachers will possess the necessary background. Rather, the overview provides examples of wisdom traditions that highlight different ways of thinking and values that can be briefly summarized and used as a point of departure for students to examine their own cultural traditions. These focused discussions will enable students to recognize the diversity of wisdom about how to nurture relationships, including how many cultures understand relationships as the primary source of wealth. Learning about key beliefs of the world's major religions/cosmologies will also lead to a broadened understanding of the many misconceptions that underlie Western attempts to integrate these cultures into the market economy that is ecologically unsustainable and to become dependent upon data as the basis of decision making about relationships.

Hopefully, the CCS reforms will be rejected. I also hope against overwhelming odds that faculty both across the disciplines and in teacher education will begin to address the cultural roots of the ecological crisis—and begin to support reforms that revitalize the community-centered and intergenerational traditions that are highlighted in the different chapters of this book, including the wisdom traditions that are being threatened by the forces promoting the digital revolution. Grassroots efforts by people who are modeling how to live less consumer-dependent lives are too many to mention here—other than to say that I am indebted to them. But I must express my thanks to Mary Katharine Bowers for her continued willingness to read yet another book manuscript and to suggest ways to bring greater clarity to what I write.

Chapter 1

How the Current Common Core Standards Reproduce the Misconceptions of the Past

At first glance, it seems that the nationwide effort to adopt the Common Core Standards (CCS) will lead to major improvements in the quality of education. What they, or some state variation, are intended to replace is the uneven quality of education students receive in public schools, which is caused by differences in the teachers' academic preparation and personal priorities for learning, home and community influences, and differences in levels of state support. As the new CCS are being represented as preparing the nation's youth for college and successful careers, they also would seem to meet the most important needs of the nation.

Indeed, this is the hope of the Bill Gates Foundation, the Business Roundtable's Education and Workforce Committee, and the Arne Duncan–led U.S. Department of Education, which has poured millions of dollars into developing and promoting this educational reform. These groups' continued belief that standardized tests provide an accurate assessment of what students have learned, and thus of teachers' performance, as well as the $350 million in federal funds being made available to develop the system of testing, will enable two powerful multistate testing consortia to exert a powerful influence over the content of the curriculum.

Given the support at the state and federal levels, as well as from the business community, questioning whether these groups' understanding of the nature and purpose of the CCS reforms will lead to greasing the slippery slope the nation is now on may appear unwarranted. Yet, the answer should be obvious to anyone who has given attention to the near weekly reports of how environmental changes are impacting daily life—from the droughts spreading across wide swaths of the country to the biblical-sized forest fires raging mostly in the West, the collapse of fisheries, and the rising temperatures that are affecting health and food production.

In addition to asking whether the CCS will provide the education needed for addressing the deepening ecological crisis, we also need to ask how these reforms will contribute to students' communicative competence in addressing the deep cultural changes resulting from the digital revolution. While teachers are supposed to have considerable leeway in choosing the content of the curriculum, the CCS reforms, and the regime of testing that could well determine both teachers' salaries and job security, there is little likelihood that students will learn about the forms of knowledge being lost as digital technologies become more widely adopted. And by not being aware of what is being lost, they will not be aware of its importance. Nor are they likely to learn about the ideology that justifies replacing workers with robots.

Also missing in their education will be an understanding of how the globalization of digital technologies and the cultural assumptions they encode will further undermine the world's diversity of intergenerational knowledge and skills that enable people to live less consumer-dependent and thus less environmentally destructive lives. This ideology, which frames understanding all activities, relationships, and Nature itself as guided by the profit-oriented values of the marketplace, is leading to environmental decisions being made on the basis of profits and losses. Using a market accounting system for determining how best to utilize natural systems raises even more complex questions about a curriculum focused on promoting analytical and communication skills that do not take account of the dominant ecological and cultural trends shaping the future.

These three forces—the ecological crisis and its cultural roots, the ways in which digital technologies are undermining intergenerationally connected communities and displacing the need for workers on a global scale, and the market-driven ideology that continues to benefit the already wealthy—while further exploiting the poor, are simply not part of the thinking of the groups promoting the national adoption of the CCS reforms. Their thinking, as well as their economic interests, are not dissimilar to those of Rex W. Tillerson, the chairman and CEO of Exxon Mobile Corporation. He is also the chairman of the Business Roundtable's Education and Workforce Committee, which is promoting CCS reforms. It is important to recall how Exxon Mobile paid scientists to give legitimacy to the claim that global warming is a hoax inspired by liberals.

The shared silence within this powerful group of promoters about the need to address the cultural roots of the ecological crisis, including how digital technologies are undermining our civil liberties and the heritage passed on through face-to-face communication, can be traced to the silences in the university education they received in the last decades of the 20th century. This silence is also shared by the educational experts who have identified the analytical and communication skills that are to be learned at each grade level.

The proponents of the CCS reforms are out of touch with the realities now experienced by the millions of workers who have already been displaced by robots (both mechanical and software programs). This can be seen in the way these reforms are being justified on the grounds that the development of analytical abilities and communication skills will prepare students for success in college and careers. This expectation does not take account of the number of American workers (somewhere between 20 and 25%) who earn less than $10 an hour. Even though these figures may change as a result of the efforts of fast-food employees to force the industry to raise their wages to $15 an hour, which will still leave them at the poverty level, any raise in the hourly wage will likely lead to replacing workers with robots. Nor does the thinking of the proponents of CCS reforms take account of how the digital revolution will continue to threaten the ability of members of the middle class to be employed in their chosen fields; this can already be seen in the rising levels of unemployment among journalists, lawyers, and academics. The last group in particular will be impacted as major universities begin offering more of their degree programs online.

The problems with the CCS reforms go much deeper than what has just been identified. They extend back into the history when influential thinkers established the convention of associating knowledge and empowerment with the explicit forms of knowledge encoded in print. This cultural tradition now carries over to what can be digitized (which also relies upon the technology of print). This reliance upon print, rather than on oral traditions, to carry forward the ideas, the historical record of achievements and wrongs done to others, and the communal wisdom has led to many benefits. But there have been losses, too, that have gone largely unnoticed—losses that will become increasingly important as the abstract patterns of thinking reinforced by print-based cultural storage and thinking further undermine the ability to recognize how behaviors, particularly those that are self-centered and consumer-driven, are deepening the ecological crisis that now is leading to shortages in potable water, sources of protein, and life-sustaining habitats.

For those who have been socialized to think of print as a source of empowerment and as one of the most important technological achievements in the shaping of human history, these brief comments on the unrecognized characteristics of print may seem a needless diversion unless they have encountered the writings of Walter Ong, Eric Havelock, and Jack Goody. A short list of the characteristics of print that are not recognized by the promoters of CCS reforms includes the following:

1. Print reinforces abstract thinking;
2. Print marginalizes awareness that words are metaphors and that they carry forward the misconceptions and silences from the past;
3. Print gives the author's cultural assumptions the appearance of representing an objective account of reality;
4. Print provides only a surface knowledge of events, ideas, and cultural processes because it cannot reproduce the tacit, contextual, multisensory embodied experiences of everyday life;
5. Print represents a world of fixed entities and casual relationships that can be "objectively" understood, and thus misrepresents how all aspects of life—from the molecular and cultural to the behavior of natural systems—involve interdependent relationships and semiotic processes that influence the continual formation of life;
6. Print is unable to provide an accurate account of the many emergent information pathways that affect relationships within both the cultural and natural ecologies.

I have discovered in the course of past efforts to explain how print reinforces a static view of the world that if the reader or listener has thought only of the many benefits of print, and not in terms of its inability to represent the complex information communicated through relationships, ongoing transformations, and processes, then identifying what is problematic about a noncritical reliance upon print will not make sense.

As will be explained later, there is no awareness on the part of those articulating the ideology that justifies the CCS reforms that their national reform effort is based on the long-held Western cultural assumptions that were formed when there was no awareness of environmental limits or the nature of cultural and natural ecologies. In short, their historically influenced mindset assumes that there is such a thing as a rational process free of cultural/linguistic influences, that there is objective knowledge, and that this is a world of discrete entities, events, ideas, and even autonomous individuals—all of which are prominent features of the CCS reforms. Also unrecognized by this mindset are the deep, largely taken-for-granted cultural patterns that are as diverse as the languages still spoken in the world. If these cultural differences are ignored, a CCS-based education will leave students unaware of when they are engaged in the colonization of other cultures—which in this age of the Internet and increasingly lethal weapons is leading to more violent forms of resistance.

Other limitations in the thinking of the promoters of CCS reforms can be traced to the shortcomings in their university education—even in elite universities. That is, while technologies now are integral and thus powerful shaping forces in nearly every aspect of daily life, with the development of nanotechnologies moving their sphere of influence into the physical body itself, there are few opportunities for public school and university students to learn about the cultural assumptions that guide the West's approach to the development of technology. Nor will they learn about the cultural transforming nature of different technologies. The claim that teachers will still have the freedom to introduce students to different curricular content is simply empty rhetoric if teachers lack knowledge of the history and cultural transforming nature of different social and mechanical technologies. Because teachers reproduce many of the shortcomings in their own educations, their students' educations will leave them less able to acquire a balanced understanding of the appropriate and inappropriate uses of technology—which requires an awareness of cultural and natural contexts. Most students graduate with the myth deeply imbedded in their consciousness that equates technological developments with progress. Thus, technologies that serve corporate interests more than the welfare of people are largely accepted as further expressions of progress. The current situation, where greater efficiencies and profits are achieved by replacing workers with robots, raises few moral concerns—except on the part of the new class of the technologically disappeared.

What are not mentioned, either in the literature that justifies the nation's need for a common core curriculum or in the extensive list of standards for all areas of the curriculum, are the lived cultural patterns that cannot be digitized or evaluated by the use of standardized tests. What cannot be digitized or evaluated by the use of standardized tests is a curriculum that engages students in thinking about environmental and social issues in terms of the different interpretative frameworks that well-educated teachers can introduce. Nor does learning to think critically about what needs to be changed and what needs to be intergenerationally renewed as the ecological/cultural crisis deepens fit with the ideology of this market-driven reform.

The promoters of CCS reforms recognize that one of the virtues of the technologies used to test what the students are learning is that it also provides a highly efficient way of monitoring a student's long-term record of achievement. The student's record can then become part of her/his digitalized identity that can be made accessible to future employers—along with a list of books and concepts that the student has encountered. In short, the CCS reforms represent the leading edge of a movement to replace teachers with computer programs. For the business-oriented groups promoting CCS reforms, using a business model for assessing the productivity of teachers will lead to the same conclusions reached by many corporations: that computer-driven machines are more reliable than workers. Just as the transition to robot-driven production reduces the power of labor unions, replacing teachers with computer programs will reduce the power of teacher unions and eliminate costs associated with retirement and health benefits—and thus reduce the cost to taxpayers.

The corporations controlling the process of testing are already limiting the teacher's curricular and pedagogical decisions to what will be tested—which will also serve as a benchmark of the teacher's effectiveness. The computer takeover of both pedagogy and the content of the curriculum also undermines what remains of local control of education. As this becomes a near total surveillance and data-recording system, the ideology that governs this emerging regime of power is closer to that of a fascist system of control than a democratic society. The control shifts to those who control the most powerful computers, which leads to the voices of the community becoming increasingly marginalized.

What is ironic is that as more grassroots groups are revitalizing democratic decision making around issues of food, community-centered businesses, uses of natural resources, and social justice issues, those in educational leadership positions who are promoting CCS reforms are undermining the small community-centered efforts being made toward learning how to live more ecologically sustainable lives. Unrecognized by the promoters of the CCS reforms is how the different global patterns are connected: changes in ocean temperatures, extreme weather conditions, disappearance of habitats and species, the amount of carbon dioxide humans release into the atmosphere, hyper-levels of consumerism, and the toxic chemicals required by the consumer-dependent lifestyle.

The earlier mention of the consciousness-shaping influence of print was not a mindless digression from the main issue of whether the CCS reforms will lead to a better prepared workforce and to more democratic decision making. The main issue continues to be whether educational reforms will begin to address how to live more ecologically informed lives, or continue to promote the environment- and community-destroying agenda of industrial/consumer-oriented culture supported by the business community that is backing the CCS reforms.

Assessing whether educational reforms such as the CCS will enable people to make the needed cultural transitions requires examining the basic misconceptions that have shaped the development of Western consciousness—misconceptions that are still promoted in public schools, universities, and the corporate-controlled media, and by the digital technologies that are now being globalized. The increasing reliance upon the Internet, which is technologically unable to represent the information-rich local cultural and natural ecologies, continues to perpetuate key characteristics of Western consciousness shaped by over two centuries of philosophers who elevated abstract thinking over the importance of learning from the intergenerational knowledge that sustained the cultural commons of different cultural groups.

The world of seemingly infinite possibilities made available by abstract thinking that lacks accountability for accurately representing local contexts, as well as by the Internet, leads to focusing attention on the immediate interest of the user and on thinking about what further explorations in cyberspace will yield. In effect, the way the Internet contributes to cultural amnesia contributes to the Enlightenment bias against the possibility that anything important can be learned from our own history or from the prehistory of people who were able to survive by giving close attention to the interdependent ecological patterns in their environments.

This mindset, which is strengthened both by the rate of technological innovation and by the emphasis on data and information that is largely context-free, will make exceedingly difficult the cultural shift from thinking of things (that is, a world of discrete objects and events) to a process/relational form of awareness. Yet this is the paradigm shift that sustainable educational reforms must address.

Clarifying the basic differences between the inherited Western assumptions about the nature of reality and the actual patterns of information exchange that occur in all relationships within both natural and cultural ecologies is essential to understanding why CCS reforms will continue to limit any genuine progress in developing an ecologically informed form of intelligence. In considering how some scientists are now understanding key features of cultural and natural ecological systems, from micro to macro systems, it is important to keep in mind that I am attempting to lay the conceptual groundwork for identifying how CCS reforms are perpetuating the most problematic and deeply held cultural assumptions that are preventing the majority of Americans who express concern about changes occurring in the environment from recognizing what has to change. And in not recognizing how the old patterns of thinking are deepening the double bind where taken-for-granted cultural patterns are accelerating the rate of environmental degradation, the proponents of CCS reforms are putting at risk the life chances of future generations.

The next chapters will focus on what some scientists and deep thinkers such as Alfred North Whitehead and Gregory Bateson, as well as a diverse group of cultural linguists and anthropological psychologists such as Walter Ong, Richard E. Nisbett, and Charlene Spretnak, are challenging us to rethink. What I have experienced in the past is that introducing ways of thinking that do not fit the taken-for-granted interpretive frameworks upon which the careers of others have been built does not lead, in most instances, to a positive response. Instead of prompting a dialog that might have pushed my own thinking in more productive directions, my efforts have led to a replication of my prairie years of experiencing the futility of pushing against the wind. And when the silence is broken, it too often is with an assertion that Bowers is again writing in a manner that is difficult to understand. That the reader's own categories may be too limited for understanding the key characteristics of cultural and natural ecologies is seldom considered. But for readers who are already aware of the experiential differences between ways of knowing that represent reality in abstract and fixed terms and ways of knowing that take account of the information pathways that are part of a world of relationships and becoming, the effort in the following chapter to clarify the basic epistemological/cultural differences that separate an ecologically sustainable paradigm from the one that we now identify with achieving material and technological progress will provide a better starting point for thinking about an ecologically informed common core curriculum.

Chapter 2

The Disconnect Between How We Think and the Relational World We Live In

Charlene Spretnak goes to the heart of the problem when she writes that "Our hypermodern societies currently possess only a kindergarten level understanding of the deeply relational nature of reality." For all our technological and intellectual achievements, we have missed, as she puts it, "the way the world works" (2011, p. 1). The challenge here is to provide a conceptual framework for understanding how the CCS reforms will continue to perpetuate students' misunderstanding of the relational way the world works. And in not understanding this, students will continue to ignore the interdependencies between natural and cultural ecologies—and thus will not know how to exercise the ecological intelligence upon which our future survival depends.

Many of the students from the early grades on, who will have learned from their Internet experiences to think beyond the limits of their own personal sense of time and space, will become increasingly alienated from the classroom as they learn the skills that fit the requirements of standardized tests. The higher drop-out rates will lead states to adopt alternative approaches to what students find a boring, repetitive, and irrelevant educational experience. Unless students are engaged in some form of mentoring programs, the alternative will likely be an online curriculum—which will be even less able to represent the relational world they live in.

Regardless of how the disconnects play out in preparing students for, on the one hand, the workplace that is becoming dominated by robots, and on the other, universities that will increasingly consist of online courses and degrees lacking both in-depth knowledge and divergent interpretations, the most basic shortcoming in the CCS reforms is that students will continue to be socialized to base their lives on the misconceptions that have guided Western thinkers for hundreds of years. In order to understand this criticism, it is first necessary to provide an overview of how the paradigm that emphasized a mechanistic view of organic processes, of individual autonomy in a human-centered world, and of science and technology leading to endless progress and material abundance is now being challenged. The primary importance of these challenges, beyond providing a more accurate understanding of life-forming processes, is that it provides the conceptual framework necessary for addressing how to live more ecologically sustainable lives.

What does Spretnak mean when she refers to the world as relational, and why does our current kindergarten understanding of how the world works become especially important as the world's population expands toward the 9 billion mark, along with a consumer lifestyle that is further undermining the life-sustaining capacity of natural systems? The answer to both questions can be traced to a single word: ***ecology***. In the middle of the 19th century this word represented what has become the modern translation of the early Greek word *oikos,* which supposedly referred to the management of the Greek household. I say "supposedly" because the translation by the German biologist Ernst Haeckel (1834–1919) was accepted as fact within the scientific community of that day. This example of metaphorical thinking, where the management of the environment was understood as like the management of the household, led to a very narrow understanding of ecology as the study of the behavior of natural systems. Lost in translation was what Haeckel, as an early proponent of Darwin's theory of evolution, was less able to understand—namely, that for the early Greeks, *oikos* encompassed the norms governing a wide range of cultural practices.

This science-dominated understanding of ecology is now beginning to change. A small group of scientists are developing the new field of biosemiotics that expands understanding of how the word *ecology* moves us closer to understanding how all aspects of the world work. There are now increasing references to the ecology of identity, the ecology of language, the ecology of bad ideas, the ecology of colonization, the ecology of marriage, and so forth. That the explanatory power of the word *ecology* can be applied to any aspect of the natural and cultural world, as well as how they interact, is based on the recognition that ecology is another word for codependent relationships and the life-shaping and multiple patterns of communication that are integral to all relationships.

This is where the thinking of Alfred North Whitehead, Gregory Bateson, the biosemiotic-oriented scientists, Charlene Spretnak, and other linguistic and anthropological thinkers such as Clifford Geertz, Walter Ong, and Richard E. Nisbett becomes helpful. Nisbett's *The Geography of Thought: How Asians and Westerners Think Differently…and Why* (2003) is especially useful as it clarifies how the languages in East Asia rooted in Confucianism, Taoism, and Buddhism focus awareness on the world of relationships and the moral codes that should guide these relationships. It is only in recent decades, however, that Western thinkers have begun to lay the conceptual foundations for understanding the misconceptions that represent the world as material entities—both animal and human—

that have their own distinct properties and that can be understood objectively and engineered to serve economic and political interests.

In Whitehead's most important and most difficult book, *Process and Reality* (1929), he challenges the idea of discrete entities or things—which range from ideas to organisms, events, and material objects, facts, etc.—by claiming that actual entities are vital, transient "drops of experience, complex, and interdependent" (p. 28). That is, contrary to the Western, linguistically-driven habit of thinking of things and objects, actual entities are units of emergent processes. As he puts it, "There is no going behind actual entities to find something more real" (pp. 27–28). In short, there are no self-contained "things," as everything in the human world has a history shaped by both environmental and cultural influences. Reality is best understood as ongoing relationships (units of process) that serve as creative influences on succeeding relationships.

It is the thinking of Gregory Bateson that brings into focus what is most distinctive about relationships, and a key characteristic of all ecologies. Bateson's *Steps to an Ecology of Mind* (1972) is also a difficult read, partly because it is a collection of essays in which his most important insights about relationships (ecologies) are only briefly explored and then submerged in a discussion of other nonlinguistic issues. If one reads him in terms of what he has say about the interconnections between the archaic language processes we still take for granted and living systems (ecologies), the pedagogical and curricular implications begin to emerge for understanding Spretnak's observation about why the high-status systems of knowledge promoted in public schools and universities, which are largely based on print-based knowledge, misrepresent how the world works.

Summaries are always dangerous, but it is possible to present Bateson's core ideas about how language encodes earlier misconceptions and silences that continue to marginalize awareness that relationships, and the information communicated through these relationships, are the dominant feature of all forms of existence. One of Bateson's criticisms of what he referred to as a recursive pattern of thinking in the West concerns the past failure to understand the individual, plant, event, and so forth in terms of its relationships within the ecological system in which it is a participant. The misconception that there are autonomous entities, and thus the ontological world created by this misconception, leads to studying their distinctive characteristics separately from the emergent life-altering relationships within the micro and macro ecologies that encompass all forms of life.

Below I discuss three of Bateson's insights about language that are particularly relevant to understanding how the current CCS reforms reinforce the long-held cultural pattern of ignoring relationships and thus the ecology of influences that carry forward a long history of previous influences. For readers who want a deeper understanding, they should go to the chapters in *Steps to an Ecology of Mind* where Bateson speaks for himself. The section titled "Epistemology and Ontology" is the most direct discussion, but other insights are scattered throughout the book. Unlike other books on the ideas of Bateson, my book *Perspectives on the Ideas of Gregory Bateson, Ecological Intelligence, and Educational Reforms* (2011) focuses on the connections among his insights on how the misconceptions encoded in the metaphorical nature of language perpetuate such myths as individual autonomy, the progressive nature of change, and the idea that science and technology will enable us to survive the destruction of the environment.

Perhaps most important is how Bateson's three core ideas on language—which are largely unknown by most public school teachers, academics, promoters of the CCS reforms, and the general public—highlight how the misconceptions about a world of facts, objective knowledge, and data help us to recognize the many ways classroom teachers and professors undermine the relational way of thinking essential to exercising ecological intelligence. These core ideas include "The Map Is Not the Territory," Double Bind Thinking and Behaviors, and "A Difference Which Makes a Difference."

The Map Is Not the Territory

As Bateson thinks ecologically, he recognizes that everything, including words, has a history shaped by earlier cultural and environmental influences. This insight immediately brings into question how the current overreliance upon print (whose limitations were identified earlier) undermines awareness of the ecology of language. The current meaning of words such as *woman*, *individualism*, *data*, and so forth is the outcome of an earlier process of metaphorical thinking where the analogs settled upon by thinkers in different cultural eras are carried forward and too often become the taken-for-granted basis of thinking about today's problems and possibilities: for example, the old analogs that framed the meaning of women have now, in some regions of the world, been replaced by new analogs that represent women as artists, astronauts, historians, CEOs of giant corporations, and so forth. The current efforts to understand individuals in terms of their relationships within the larger ecologies they are dependent upon represent the current effort to reframe how individualism is understood. Other cultures have already achieved a relational/ ecological way of thinking about the individual, while others continue to derive their analogs from archaic meta-narratives.

The important issue here is how old patterns of thinking misrepresent today's realities. They continue to be based on the root metaphors (interpretative frameworks) of patriarchy (now being challenged), individualism, progress, mechanism, a human-centered world, economism, and now evolution, that go back hundreds of years—and in the case of patriarchy and anthropocentrism (human-centeredness), thousands of years. One of the characteristics of root metaphors is that they create supporting vocabularies that make it difficult to challenge what the root metaphor or combination of root metaphors excludes from awareness.

For example, the vocabulary that supports the root metaphor of individualism, such as *freedom* and *autonomy*, limits the possibility of recognizing that words have a history, and that many of the individual's taken-for-granted patterns of thinking are based on the metaphors that encode the assumptions from earlier eras. In effect, the relational nature of what is mistakenly thought of as the autonomous individual needs to take account of how her/his patterns of thinking, personal identity, and even physical characteristics have been influenced by the ecologies of language, cultural identity, and genetic inheritance. The root metaphor of mechanism, which can be traced back to the thinking of 17th-century scientists such as Johannes Kepler, led to a vocabulary that is now used to explain organic processes, including the nature of thought itself. Other root metaphors such as evolution and progress have also led to complex vocabularies that are self-reinforcing of its deepest conceptual foundations. The excluded vocabularies limit awareness of other relationships that, as the ecological crisis deepens, are especially critical to achieving a sustainable future.

If students are to learn to think relationally, which is needed for developing an ecological understanding of the world they live in, it is important for them to be introduced to Bateson's explanation of an aspect of language that has generally been ignored—that is, his explanation of what I prefer to call the linguistic colonization of the present by the past. The metaphor of "**map**," as he uses it, refers to the conceptual interpretative framework that is based on the vocabularies (metaphors) acquired in becoming a member of a language community. The "**territory**," for Bateson, refers to the current everyday world of relationships—that is, the cultural and environmental ecologies within which we live. In short, the maps (the metaphorically constructed interpretive frameworks) are generally inadequate guides for understanding and responding to current social and environmental changes. This is because the selection of analogs in the distant past, such as thinking of the environment as a source of danger and in need of being brought under human control, and then later as a natural "resource" waiting to be economically exploited, was not based on an awareness of the interdependencies between the natural and cultural ecologies. The root metaphors of mechanism and progress, which provided conceptual direction and moral legitimacy to the early stages of the scientific/industrial revolution, also limited awareness of the exhaustible nature of natural resources.

We shall later consider how students can be mentored in becoming aware of how the metaphorical nature of language illuminates or hides an awareness of what is communicated through their relationships with each other, of the traditions from the past still carried forward in their behavior and values, and of the natural systems undergoing changes that exceed the capacity of technology and science to reverse. This will be taken up when considering how the CCS reforms misrepresent the ecology of language.

Double Bind Thinking and Behaviors

Double binds were first understood by Bateson and his followers within the context of therapy situations where the efforts to help took the form of reinforcing the very behaviors that needed to be changed—thus making the idea of progress an illusion. But the concept has more important implications in terms of understanding the double bind inherent in current widely held cultural agendas such as the globalization of the West's economic system, and of digital technologies, and in the use of the English language that privileges nouns over verbs—to cite just three examples of double bind thinking.

The linear view of progress taken for granted by the promoters of world economic growth fails to take account of environmental limits—thus the double bind thinking leads to equating with progress the economic exploitation of the whole biosphere we depend upon. The double bind in promoting digital technologies on a global basis is that this view of progress undermines the oral traditions essential to the intergenerational renewal of the cultural commons that enable people to live more community-centered and thus interdependent lives that rely less on consumerism. Double bind thinking results from relying upon the old assumptions (conceptual maps) instead of giving attention to what is being communicated in relationships that have a smaller ecological footprint.

The double bind in the process of linguistic colonization where English displaces other languages is that English nouns such as *individualism, progress, intelligence, facts, environment,* and so forth reinforce a world of fixed entities that seemingly are independent of actual cultural contexts and the ecologies of relationships. That is, they reproduce a static view of reality, rather than the relational/process/emergent world communicated through the use of verbs. Linguistic colonization of other cultures can be seen in how the adoption of the English vocabulary that now accompanies Western technology and consumerism within East Asian cultures, along with the printed texts of Internet technologies, are undermining their more relationally sensitive languages. As in the earlier examples, double bind thinking fails to recognize that what is assumed to be a progressive development is in reality a destructive set of ideas and practices. Unfortunately, the language that accompanies double bind thinking, and appears essential to a modern way of thinking, hides its own history of failure in solving fundamental social and environmental problems.

"A Difference Which Makes a Difference"

This phrase is part of Bateson's statement on what occurs in relationships. As it is a key to understanding both what he means by double bind thinking and how the conceptual maps are seldom adequate guides to understanding and responding to today's "territory," it is important to quote him in full. "A 'bit' of information," he writes, "is definable as a difference which makes a difference. Such a difference, as it travels and undergoes successive transformations in a circuit, is an elementary idea" (1972, p. 315). Bateson follows this brief statement with the example of the series of differences which make a difference, such as how the cut-face caused by the axe introduces a difference that leads in turn to a change in the angle of the axe as it makes the next cut. The response of the Other to the difference which makes a difference can be observed in every relationship—in speaking with others, in playing a game, in walking through a forest, in exploiting someone else, and so forth. As Foucault put it, an action leads to a responding action. Even refusing to respond is an action or difference that leads to a difference in the relationship—which may cease to exist.

His brief statement and equally brief example are not really adequate for overcoming how we have been conditioned to think of acting on things, and to ignoring how we continually adjust our response to the difference which makes a difference in making bread, in playing a game of chess, in conversing with others, in passing another car, in supporting the clear-cutting of an old-growth forest, in driving a car that every year puts 8,320 pounds of carbon dioxide into the atmosphere, in being passive as computers replace workers and further erode our privacy, and so forth. These examples of relationships encompass both cultural and natural ecologies, as well as the micro and macro scales of these interacting ecologies. And there is no escaping from them. The question is whether we can become aware of the historical influences that limit our awareness. Also, can we become aware of the ecological destructiveness of the old conceptual/cultural maps that represent individuals as rational and autonomous, and which act on the external animate and inanimate worlds?

The reality is that we all adjust our thoughts and behaviors to the differences that our language and personal sense of awareness enable us to recognize as we interact in the complex ecologies that are an inescapable aspect of daily life. Our response to some of the information Bateson refers to as "differences" is greatly influenced by the conceptual maps (metaphorical language) framed by earlier thinkers. The overreliance upon print and digital technologies continually reduces the emergent world to things, events, facts, and static relationships. The metaphorical nature of language, with its historically derived analogs that frame how to interpret the world in terms of past ways of thinking, hides not only the interactive processes that are part of our living world, but also what earlier thinkers were unaware of. The culturally influenced sense of being an autonomous individual, with an inflated sense of personal agency and privilege, also leads to a reduced awareness of what is being communicated through the multiple information pathways that are part of even the seemingly most banal relationships.

Let me cite two examples of seemingly simple relationships that turn out to be complex in terms of the different kinds of information that are being communicated—but mostly ignored because of cultural influences such as biases, lack of sensitivity and empathy, and the personal egos that the participants bring to the relationship.

First, it is necessary to clarify a potential source of confusion. I have been using two metaphors, *information* and *communication,* which are hangovers from the old paradigm that represented the world as a distinct entity and the individual as a rational being who supposedly can provide an objective account of her/his observations of the external world. Bateson's reference to "differences which make a difference" needs to be understood as involving different messaging systems (or "information") that may include the electrical/chemical, genetic, differences in temperature, and so forth that influence what cells communicate to each other—and which may inhibit or promote growth. The complex physical/chemical changes in one's own bodily experience may become a part of the differences (information) which make a difference in how one responds when encountering someone where tensions still exist. The connections between systems and what is communicated between them was highlighted when the 2013 Nobel Prize in medicine was given to three researchers who discovered how hormones inside a cell that are ferried in membrane-bound sacs known as vesicles know how and where to deliver their genetic information so that there are no disruptions that can lead to a wide range of physical ailments. The complexity of the information exchanged, for example, can be seen in how the molecular code carried in the vesicle senses calcium ions and triggers the release of brain chemicals at the right time.

In terms of cultural patterns of communication, the range of "information" generally includes nonverbal cues that send powerful messages about how the relationship is interpreted, as well as the use of words (metaphors) and silences that convey historically loaded prejudices and so forth. For example, when I tried to talk to colleagues from other academic departments about the importance of the cultural commons, the differences which made a difference for me were communicated in how quickly they avoided eye contact, changed the subject, and signaled with bodily movements that they needed to go elsewhere. These differences in behavior, like all relationships, need to be understood as ecologies that were influenced by the professor's conceptual background—including influences that contributed to her/his being curious about a new way of thinking, or defensive in protecting a self-image of being a leading thinker. And these ecologies also include the ecology of language that limits or involves an expanded vocabulary necessary for understanding newly encountered ideas. The ecology of thinking within the professor's discipline, as well as the ecology of values and reward system within the department and within the discipline at the national and even international levels, all influence the professor's response to what was being communicated in the short-lived relationship.

Something more needs to be said about understanding relationships, whether at the scale of nano or macro ecological systems. That is, relationships should be understood as information pathways within and between the relationships within the larger ecologies. Communication varies in terms of the Other's semiotic patterns, which for humans may be limited by the metaphorical language inherited from the past.

At some point, teachers should challenge the archaic idea that we exist as autonomous beings in a world of material and unintelligent things by introducing students to Bateson's insight that relationships are ecologies of differences that lead to reciprocal responses—in effect, a dance of information exchanges that influence subsequent behaviors. Students could be asked to observe the nonverbal patterns, as well as the changes in the use of language, that are part of every conversation and relationship. The interactive world that Bateson's phrase highlights can be seen in the differences in the behavior of a pollinating insect flying around a nonnative plant. Students should be asked to give special attention to the difference which makes a difference in the behavior of the insect. That is, what are the sources of information to which the insect responds? Do the past influences include the genetic makeup of the insect as well as the plant? Why do so many people want to rid their yards of native plants? Does the absence of native plants have any relationship with the decline in the number of pollinators? How do the chemicals in the soil become critical differences which make a difference in the growth of the native flowers to which the insect responds?

Similar everyday examples, such as sporting events, conversations—including those between people of different genders, social classes, and ethnic groups—learning from others how to plant a garden or engage in a craft, and so forth, can be used to encourage students to give close attention to the differences (information) communicated as the dance of relationships evolves.

An example that will engage the students' attention, as well as make explicit the ecology of differences that comes into play in even the most banal relationships, was suggested by Clifford Geertz. In his explanation of "thick description" (1972), which is really what is being suggested here as learning to give explicit attention to the differences which make a difference (including historical and otherwise taken-for-granted patterns of influence), Geertz suggested that his readers consider what separates an involuntary wink of the eye from the wink that is intended to send a message to another person. What then are the differences that might influence how the intended wink is understood and responded to, or behaviors that follow from a series of misunderstandings? What are the behavioral and other changes occurring in the local context? How does memory influence how the relationships prompted by the wink will evolve—and even be misunderstood? How do gender and social status differences become part of the message exchange?

Another common everyday relationship that involves multiple messages that can lead to misunderstanding, depending on the taken-for-granted largely influenced cultural assumptions the participants bring to the relationship, is the way people engage in different forms of physical contact. The growing tendency toward engaging in physical embraces is an example of ecologically complex messages—that is, differences that should have made a difference, where what is ignored could become a new set of differences that become part of a new succession of differences that undergo "transformations in a circuit" (to get back to Bateson's wording). Having students observe how and when people embrace each other, as well as the nonverbal patterns of communication that follow the embrace, provide yet another example of the complex range of transformation in the differences which make a difference. It will also provide a good example of what Spretnak and others are saying about living in a world of relationships—and awareness that may lead to reducing the mindless behaviors that set off a string of consequences that go unnoticed when the complexities of relationships are ignored.

The small group of scientists who were influenced by the ideas of Bateson as well as others such as Thomas Sebeok who focused on the ecology of communication among animals, and by the growing body of research on how cells interact, are now promoting biosemiotics as a way of understanding the relational life-forming and -sustaining (and -destroying) processes. Professor Shih-yu Kuo, who recently retired as a professor from a Taiwan university and from the Institute of European and American Studies, suggested that this new field of inquiry should be called "ecosemiotics" if the study of culture is not to be overshadowed by the continuing emphasis on the natural sciences. Her suggestion leads to the more inclusive understanding that all relationships, in both the natural and cultural worlds, involve some form of semiotic (information) exchange that sets in motion further exchanges.

Jesper Hoffmeyer, the Danish molecular biologist who is one of the leading thinkers in this emergent field of inquiry, reframed Bateson's statement about differences which make a difference being an elementary idea by suggesting that the multiple forms of information communicated through differences should be understood as signs. He further shifts the focus from the traditional mechanistic way of understanding the primary characteristics of things, plants, animals, cells, and so forth to what is occurring in their relationships. This can be seen in Hoffmeyer's observation that "the individuality of a human life cannot be justified by its uniqueness as a particular genetic combination, but must be justified by its uniqueness as a particular semiotic creature" (2008, p. 328). Thus, the individual, for example, is not to be understood only as having the capacity of being intelligent and a critical thinker, of being ego-centered, hard-working, and so forth. Instead of focusing on the personal attributes that might be identified by liberals and theologians, or by teachers, Hoffmeyer suggests that the focus needs to shift to the biological and cultural attributes that enable participation in different semiotic systems of communication. For example, humans lack the genetic and cultural attributes that enable them to respond to the signs that enable dogs to recognize dangerous substances. Nor are the semiotic systems that orca whales rely upon available to humans, given their differences in genetic and cultural makeup. In short, Hoffmeyer is shifting the focus from the narrow range of communication that educators and others too often associate with speaking and writing to include the whole range of life-forming processes—from the most primitive to the most complex and evolved organisms.

In introducing the idea that a more complex interspecies understanding of communication requires shifting to the more inclusive category of signs and semiotic systems that all organisms (including humans) have the genetic and culturally mediated capacity to respond to in terms of their unique form of agency, Hoffmeyer and the others in this new field have provided a way of understanding what Bateson meant by writing that differences (which is the most basic form of communication within ongoing ecological life-altering processes) represent the most basic idea or unit of information. In effect, biosemiotics (or "ecosemiotics," as I prefer) is in the Whitehead and Bateson tradition of representing reality as emergent and ongoing processes. What it adds is an evolutionary framework, and a way of understanding that the biological and cultural worlds represent different levels and forms of cognition (that is, the ability to respond to signs) at even the most primitive level.

As most people have been socialized to take for granted that they act on a world of things, and should strive to understand how to achieve greater efficiency and power in the use of things while accepting the silences in the conceptual and moral heritage of previous generations, it may be difficult at first to recognize what is ecologically problematic about CCS reforms. The following chapter will make explicit what Bateson refers to as the recursive patterns of thinking reinforced in how the teacher is to guide the acquisitions of subject matter and skills in the areas of language, history, social studies, and the sciences and technology. Chapter 4 will address the silences in the CCS reforms that are reproduced both in the work settings that will be increasingly driven by computers and in most university classrooms. The silences that chapter 4 will focus on include the ecologically and thus community-sustainable forms of knowledge that are dependent upon mutually supportive relationships—that is, the cultural commons now being threatened by the further expansion of a market economy and by the growing dominance of digital technologies as sources of cultural storage and thinking.

Contrary to the widely held misconception that individuals, either through a rational process or by subjective whim, decide what constitutes the moral values that will be appropriate in different relationships, chapter 5 will address what is totally ignored in the CCS reforms: namely, today's need for basing relationships with each other and with the environment on different traditions of wisdom. Given that leading computer scientists are now claiming that super-intelligent computers are close—indeed, within a few decades of surpassing human intelligence—introducing the word *wisdom* may seem pointless. However, as technology and the market system are fast reaching a tipping point where poverty and hopelessness will transform today's passive acceptance and mild street demonstrations into more violent actions that will shake the foundations of social life, what will be needed in rectifying the extreme inequalities that separate the vast number of impoverished people from the wealth and power of the small elite that now controls the political process is *wisdom*—a word that should have a more prominent place in today's political and educational vocabulary.

Chapter 3

Behind the Appearances: Conceptual and Moral Double Binds Inherent in CCS Reforms

The most relevant question to ask about the CCS reforms is whether they will enable students to understand that they live in a world of ecological relationships. The current justification for the CCS reforms, which focus on employability and achieving success at a university that will lead to a career, makes the above question appear strangely out of touch with today's economic realities and the market liberal ideology that drives them. Yet, the question goes to the deepest conceptual foundations of Western culture—including the issue of whether society will be able to make the transition to a less ecologically destructive lifestyle. As suggested in the last chapter, the key criterion for judging the merits of CCS reforms is whether students will acquire the ability to contribute to life-enhancing and -sustaining relationships by being able to recognize what is being communicated through the information pathways that make up the larger cultural and natural ecological systems upon which they are dependent.

If we give close attention to the justifications for specifying a common set of analytical abilities and skills that will be tested at each grade level, we find what Bateson identified as double bind thinking. That is, double bind thinking occurs in situations where well-intentioned ideas and agendas, instead of overcoming the current social and ecological problems, actually add to them. To recall an earlier example, double bind thinking occurs when, in being aware that human behavior is changing the world's chemistry and thus life-supporting process, we continue to promote unlimited economic growth and an unrestrained consumer lifestyle. Another example is students who take on massive debt to obtain a university degree while knowing the unlikelihood of obtaining a job that will enable them to pay off the debt. Albert Einstein summed up the key feature of double bind thinking when he said, in essence, that we cannot fix a problem by relying upon the same mindset that created it.

The question not being raised by the promoters of the CCS reforms, and by the many school administrators and classroom teachers who are preparing to implement them, is whether the basic cultural assumptions that underlie the reforms are essentially the same as those that provided conceptual and moral legitimacy to the industrial revolution that is now entering the digital phase of globalization. These assumptions have led to a culture in crisis, with massive maldistribution of wealth, a lifestyle that depends upon exploiting the world's natural resources as rapidly as possible, and, most importantly, the breakdown of our democratic traditions. There is no mention of these problems by the promoters of the CCS reforms. Instead, they hold out the promise that students will become "engaged and open-minded—but discerning—readers and listeners," that they will be able to "cite specific evidence when offering an oral or written interpretation of a text," and that they will learn that …"in the 21st-century workplace they will need to work together with others of different cultural backgrounds" (*Common Core State Standards Initiative,* 1990, p. 7).

The emphasis on acquiring these personal qualities hides a double bind that the promoters in the business community want to ignore, especially since one of the cultural assumptions they take for granted is that the welfare of the worker must always be subordinated to achieving greater efficiencies and profits—which increasingly means replacing workers with computer-driven machines. The promise that the CCS reforms will enable students to become "self-directed" learners in a world increasingly controlled by computers being engineered to move us into the era of singularity, where computer intelligence will surpass human intelligence, suggests that the promoters of the CCS reforms are unaware of what prominent computer scientists envision for the rest of humanity (Moravec, 1990; Stock, 1993; Kurzweil, 2005). The goals of promoting open-minded learners, of learning to thoughtfully employ technologies, and of becoming self-directed learners—goals used to justify earlier educational reforms—are couched in empty rhetoric that hides the cultural assumptions reinforced by CCS reforms.

It is necessary to challenge the most basic misconceptions surrounding how teaching and learning are being represented. It is especially important to understand these misconceptions when considering the role that testing will play in assessing the teacher's performance and what the students are expected to learn.

Abstractions that Obscure the Complexity of Teaching and Learning

The basic misconceptions about the processes of teaching and learning can be attributed to the way print provides a simplistic understanding of the relational world we live in—including the complexity of teaching and learning. The Introduction to the *Common Core State Standards Initiative* provides a good example of the shortcomings of print. Both teachers and students are represented as abstract entities, as if there were such a human being who is free of the ongoing influences of the multiple cultural and biological patterns of information exchanges best understood as the cultural and environmental ecologies within which they are nested. To quote from the Introduction, "The standards define what all students are expected to know and to do, and not how teachers should teach" (2010, p. 6). This statement suggests that students are to be thought of as a class of individuals who are free of cultural/linguistic influences.

The reality is much different. The relational world of the student may include coming to school hungry, angry, and confused about conflicts in the home, wondering why a friend failed to respond to a text message or what has been put up on Facebook, taking for granted that what is important in life is wearing the right clothes and being popular, being aware that one is not a member of the elite group in the school, being concerned with avoiding cliques that are headed toward trouble, lacking self-confidence and having the limited vocabulary that results from parental indifference, and so forth. Not only is there an ecology of earlier influences that range from ethnicity, gender, social class, home conflicts, and sources of motivation to personal concerns about appearance vis-à-vis group norms and diet-related issues, but also there are everyday experiences with parents and classmates that may distract or help focus the student's attention and willingness to make the effort to learn what seems to be unrelated to the student's interest and experience.

The student's understanding of the choices in her/his relational world (ecologies) is influenced by memories as well as what is hoped for in terms of the future. And what are seen as possibilities always involve choices about what will be prioritized—and thus how well the student will perform when being tested. The tests will have been created by people who are unaware of the psychological/cultural influences on the student's performance.

The thinking that goes into the construction of the test will focus on the skills and knowledge that are equally abstract—such as "by the end of grade 12, [students will be able to] read and comprehend science/technology texts..." (2010, p. 62). Perhaps the text was written by scientists such as E. O. Wilson or Richard Dawkins. Both cross over into the realm of scientism when making claims about culture for which there is no scientific evidence. Or perhaps the technology text was written by Hans Moravec or Ray Kurzweil. Both argue that super-intelligent computers will soon replace humans in the process of evolution. Will the teacher of this part of the curriculum be able to help students recognize when the scientific and technological writings of these authors drift into the area of scientism? How will a test be able to assess the student's comprehension of a science or technology text? Is the ecology of the text to be taken into account, such as the dominant scientific discourse that is being challenged or supported, and the cultural/linguistic influences on the scientist who wrote the text—including the silences in her/his education, and so forth?

Just as objective facts, data, and information exist only in the realm of abstractions that supposedly are sealed off from cultural/linguistic/ego-personality influences, the technology of testing is unable to take account of the ecologies of relational knowledge and influences on the thinking of the people who create the tests, the people who evaluate the test results, and the administrators who make the final decision about what the performance of the students means when judging the teacher's effectiveness. To make this point in a different way: The experts who identify the learning objectives, and create and evaluate the test results, were educated in the last decades of the 20th century, and thus acquired conceptual maps that did not take account of the ecological challenges of the 21st century. Their abstract and now archaic and even reactionary patterns of thinking will also prevent them from recognizing that the test results do not take into account that the teacher may have been given a class of students whose abilities were limited by poor teaching in the earlier grades, as well as by chaotic home environments and poor diets. Test results, like English nouns, marginalize understanding the contextual and relational world of the students, teachers, and even the politicians and corporate leaders promoting the CCS reforms.

This is evident in how teachers are represented as abstract entities who are to make decisions about the curriculum in addition to the specific content that will be tested. Nevertheless, it is the data that counts in assessing the performance of the students, and this will become the basis for assessing the teacher's ability. The personal life of the teacher, including his/her background education, self-concept, interest in challenging students versus just putting in the time, shifting moods as teaching is reduced to a lock-step curriculum, and methods for dealing with students addicted to cyberspace communication, also come into play every hour in the classroom. Again, the teacher who is free of all psychological and cultural/linguistic influences (both past and present) exists only in the fictional world of print.

Not all states have adopted the Common Core Standards, and some states have adopted only certain segments such as the science and technology standards. Regardless of which standards are supported by the business and educational leaders in every state, as well as every other education-minded group that does not subscribe to the nihilistic idea that students should determine what they want to learn, they all can support the general goals used to justify the CCS reforms.

The following list of goals also fails to acknowledge the complexity of social and environmental issues that are leading to widespread disagreements about what constitutes evidence, or whether it is even possible to engage in an exchange of ideas with those who are entrenched in outmoded ideological positions. It is also important to note that the CCS educational goals represent a level of idealism that far surpasses the daily practices of classroom teachers, professors, and the general public. It is important to keep the following list of goals in mind, as they are stated in such generalities that there is no possibility of using a test to verify that they have been achieved. This will leave the door open for powerful groups to justify a continuation in the "revolution" in how public school and university education is to be "delivered" in the coming years (for the sake of clarity, I substitute "Students" for "They" in the following extract):

> Students demonstrate independence; Students build strong content knowledge; Students respond to the varying demands of audience, task, purpose, and discipline; Students comprehend as well as critique; Students value evidence; Students use technology and digital media strategically and capably; Students come to understand other perspectives and cultures. (*Common Core State Standards Initiative*, 2010, p. 10)

The paragraph-long explanation of the intellectual qualities associated with each of these goals fails to acknowledge the complexity of the personal and social issues that accompany every skill, fragmented bit of knowledge, and relationship. The two following goals are typical of the reductionist thinking reinforced throughout the standards upon which the student is to be tested. Students, according to one of the standards, must be "able to communicate effectively with people of varied backgrounds." A second standard requires that students "will be able to "adapt their communication in relation to the audience, task, purpose and discipline." The generality of the goals serves several useful purposes. First, like so many high-sounding statements that few people will disagree with, they provide a basis for diverse social groups to come together in supporting the general idea of a common core of knowledge that should be shared across the country. Second, they also provide cover for powerful groups to frame the educational reforms in ways that serve their specific economic and political interests. Various social groups would oppose these reforms if it were more widely recognized that powerful business groups and their political allies are behind these reforms. For example, if environmentalists were aware of how the CCS reforms will become yet another obstacle to educational reforms that enable students to learn how to live less consumer-driven lives and to become more ecologically responsible citizens, they would be raising public awareness about why the reforms should be challenged.

The increased drop-out rates of the cyber-generation students, the falling test scores, and the graduates of the CCS reforms who lack the motivation and skills required in the increasingly automated workplace will provide the "evidence" needed to justify substituting more software programs for classroom teachers. Similarly, universities, in spite of their lofty rhetoric about how the face-to-face encounter with the minds of their professors is a life-altering and character-shaping experience, will find that the economic advantages of offering more of their courses and degrees online to an international marketplace of students is far more important than their rhetoric, which is already strangely out of date in this digital/surveillance/market-oriented world. The business supporters, as well as the politicians and educational reformers always in search of their own relevance, will be nearly 100% behind making the transition to computer-mediated learning the basis of every facet of the curriculum.

Replacing teachers with computers would do away with the unpredictability of the teacher's personality, level of motivation, and intellectual background, which is often dated—and most importantly, it would do away with teachers' unions. From the perspective of a majority of Americans, the real advantage of online education is that it reduces the cost of teachers' salaries and other benefits, which do not have to be paid when relying upon computer software programs that can be used over and over again. There is also the benefit of being able to avoid the cost of new buildings. The ideological basis of support for replacing teachers with computers is the current market-driven progressive ethos that has little regard for the fate of people who are losing their source of employment and professional careers to the spread of digital technologies. Blue- and white-collar workers, journalists, lawyers, professors, and now even anesthesiologists are being replaced by computer software programs. Do teachers think the digital revolution will bypass them? Are teachers aware of how the progressive cultural assumptions they reinforced in classrooms, as well as their silences about the cultural non-neutrality of technologies, are partly responsible for the threat that will shortly overtake them?

How the Common Core Standards Reproduce the Conceptual Errors of the Past

If we can momentarily escape from how we have been socialized to think of the world in terms of discrete objects that have certain properties that will interact in predictable ways if mechanically, chemically, or mentally re-engineered, there is a greater possibility of recognizing that the biosemiotic scientists, Whitehead, Bateson, and the others who are explaining the relational world we live in have the more accurate understanding of life-forming and -sustaining processes. Many people are becoming aware of the importance of local contexts in making political decisions, in discovering personal talents and interests, in learning to live less toxic and consumer-dependent lives, and in becoming aware of the heritage of social justice thinking that needs to be intergenerationally renewed. These people represent the 21st-century citizens who are working to slow the rate of environmental degradation that threatens the prospects of future generations. These grassroots-oriented people are aware of the relational world in which they live, which is profoundly different from the understanding of most consumer-dependent people who do not ask about the adverse environmental and social justice impact of a culture that exploits hundreds of thousands of assembly line workers, or levels mountains in order to extract the coal, or dumps billions of tons of carbon dioxide into the atmosphere, increasing the acidification and warming of the world's oceans.

It is important to note that none of the thinking that is ecologically oriented toward strengthening the self-sufficiency of local communities and addressing the major crisis that will only deepen in the decades ahead is mentioned in the CCS reforms. While there is an increasing number of community-oriented businesses and business groups that are focused on reducing their ecological footprint, such as the Business Alliance for Local Living Communities, which represents 30,000 community-centered businesses in the United States and Canada, the majority of large businesses (and corporations) that want a CSS-educated workforce use the language of sustainability to justify the further expansion into global markets. They also continue to view the ecological crisis as nonexistent. And if climate changes are acknowledged, they are explained as natural and thus nonhuman-caused changes in the earth's historical cycles. Their thinking is clearly evident in the silences in the guidelines of the CCS reforms.

Making the transition to a relationally oriented form of consciousness and self-identity will be especially difficult for students who have been indoctrinated into thinking within the old paradigm that has dominated the West for over the last five or so centuries—with many of the most dominant conceptual traits extending back for thousands of years. That many indigenous as well as East Asian cultures have over the centuries developed a relational and interdependent form of consciousness has been documented by such scholars as Frédérique Apffel-Marglin, Rupert Ross, Keith Basso, David Suzuki and Peter Knudtson, Donald Munro, and Richard E. Nisbett, among others. What is being suggested by the deep study of many non-Western cultures as well as by the Western relational thinkers such as the biosemiotic scientists,Whitehead, Bateson, and Spretnak is that understanding that all life exists in interdependent relationships with other forms of life is not a new discovery whose implications for daily life are not yet understood. Rather, we all take into account information and other culturally mediated signs communicated through our social and environmental relationships. The problem is that most Americans have been socialized to think of themselves as autonomous observers of an external world where only the relationships that advance self-interest are to be taken seriously.

The real question is whether the awareness of what is communicated through relationships goes beyond self-interest to include an awareness of how the ideas and values we act upon contribute to strengthening relationships within the community and within the natural systems—or if awareness is limited because of the misconceptions and silences promoted through the various avenues of socialization such as schools, the media, shopping malls, and the workplace.

Curricular Examples that Marginalize Awareness of the Relational World of Cultural Interpretations and Contexts

To understand how the analytical and communication skills as well as the fragments of knowledge promoted by the CCS reforms reproduce the old patterns of thinking that can be attributed partly to print-based cultural storage and patterns of thinking, we need only examine examples of what the students will be tested on. The examples reflect what is to be learned at the different grade levels. One of the important characteristics of how the concepts and analytical processes are listed is that there is no hint that teachers will bring their own judgments to what needs to be stressed, or that they will use examples to strengthen the students' understanding. The list of concepts and analytical and communication skills reads like the list of ingredients found in a cookbook. Like the subjective decisions of the cook standing over a hot stove, the teacher's determinations about what should be stressed or ignored are highly subjective—which is what the replacement of the teacher by a computer software program would seemingly eliminate. But it would not really. The subjectivities and misconceptions of the people who write the software program simply move the element of subjectivity that surrounds what is to be learned and tested further into the black and unaccountable hole of digital culture.

Examples from Various Grade Levels

Under "Reading Standards for Literature" in *Common Core State Standards Initiative* (2010), we find the following:

Grade 4: Determine the meaning of words and phrases as they are used in a text, including those that allude to significant characters found in mythologies. (p. 12)

The question the teacher is not likely to ask is whether words have universal meanings—that is, the same meaning in all cultures—and if their meanings have changed over time within the student's own culture. These are issues relating to linguistic relationships, including the complex ecology of cultural relationships that influenced the choice of analogs that frame the meaning of words—which are mostly metaphors. The most basic misconception that students will learn is to take for granted the conduit view of language (the sender/receiver view of language) that hides that words have a history and that their currently accepted analogs that frame the meaning of words (metaphors) reproduce the misconceptions and silences of earlier cultural eras. What teachers will fail to explain to students is that the conduit view of language is essential to maintaining the myth of objective facts, information, and data. What is to be learned in grade 4 about the meaning of words is to be covered again in grades 11 and 12. Again, there is no recognition that students will not learn about the historical role of metaphors in reproducing in current thinking the misconceptions of the past. Indeed, if the heads of corporations and politicians supporting the CCS reforms understood that learning the meaning of words should also require learning about the history of metaphorical thinking, which can be better understood both as the linguistic colonization of the present by the past and as the linguistic colonization of other cultures, they would be critics rather than supporters of the CCS reforms.

Grades 6–8: Distinguish among facts, opinions, and reasoned judgment in a text. (p. 61)

This is a good example of how a nonrelational understanding of the world is perpetuated in public schools and universities. The question that students will not be asked to consider is, does an opinion take on the appearance of a "fact" when it is put in print, or asserted as a fact by the speaker? This is no minor issue. Should teachers be reinforcing the idea that the lack of contexts, the historical influences on language, and the role that print plays in creating the myth of objective knowledge are unimportant? The misconceptions reinforced by what students will be tested on will play a fundamental role later in life when politicians and the people promoting replacing workers with machines claim that the "data" and "objective information" provide evidence that robots are more efficient and cost-effective (that is, more profitable) than humans. Reinforcing the myth of objective knowledge and facts is crucial to hiding the political nature of languaging processes—which is exactly what undermines the democratic process.

Grade 8: Evaluate the advantages and disadvantages of using different mediums (e.g., print or digital texts, video, multimedia) to present a particular topic or idea. (p. 39)

How many teachers have the conceptual background necessary for helping students understand the differences between oral and print-based cultural storage and thinking? And how many understand the forms of knowledge, especially about the relational world we live in, that cannot be reproduced by digital technologies? A basic aspect of human relationships is the use of memory. The question that few teachers are likely to ask is whether any of these technologies accurately reproduce the contexts that historically shaped personal and community memories. That is, do they contribute to the cultural amnesia that is such a dominant feature of modernity, where the history of social justice movements and the legacy of cultural commons knowledge and skills are no longer relevant to the cyberspace generation? Few teachers will ask how these technologies contribute to undermining the intergenerational knowledge of other cultures and the student's own cultural commons, which will become increasingly important as digital technologies further reduce the opportunity to earn a living.

Writing Standards, Grades 11–12: Write arguments to support claims in an analysis of substantive topics or texts, using valid reasoning and relevant and sufficient evidence. a. Introduce precise, knowledgeable claim(s), establish the significance of the claim(s), distinguish the claim(s) from alterna-

tive or opposing claims, and create an organization that logically sequences claim(s), counterclaims, reasons, and evidence. (p. 45)

On the surface, this sounds like a worthwhile educational exercise. Motivating students to give close attention to what they read, and to follow the development of an argument and give attention to what constitutes evidence for various claims, is indeed a genuine achievement in this era of cyberspace-induced short attention spans and the constant need to be entertained. However, the educational merits are severely compromised by the failure of the educators who wrote this standard to understand that learning often involves mentoring relationships—especially when it comes to the cultural complexity of human relationships. A supposedly rational process that is free of cultural/linguistic influences, including ideologies, has been the basis for exploiting both the environment and other peoples. That is, even the most exploitative behaviors are justified on rational and logical grounds. Rationality and logic are two of the most misused, context-free terms in our vocabulary.

The question that the curriculum experts who wrote this standard failed to ask is whether the student needs to understand the different belief systems (which may include ideologies such as libertarianism, conservatism in the Edmund Burke/Wendell Berry tradition, and various expressions of liberalism, as well as religious belief systems) in order to know what constitutes "relevant and sufficient evidence." Should the student be able to understand the different belief systems—and the influence of the vocabularies that support these belief systems—while excluding understanding other ways of knowing? Should students also understand the cultural assumptions that underlie what they regard as the basis of their "own" rational process? These are questions about the relationships between cultural/linguistic influences and the student's own supposedly rational process—including decisions about what constitutes evidence—that will be taken seriously by the other participants in the relationship.

This standard also raises questions about how the student's performance will be evaluated. Will it be evaluated by a software program that encodes the programmer's cultural assumptions and lack of knowledge of different ideologies and other belief systems? Would students from different cultures respond differently to what constitutes knowledgeable claims and supporting evidence? Again, critically important relationships come into play when evaluating the student's performance. The classroom teacher's and programmer's depth of background knowledge, as well as prior socialization to their cultural group's taken-for-granted beliefs and values, awareness of cultural differences in ways of knowing, and the reward system for forcing the complexities of life to fit what is required by the technology, are always part of the ecology of testing.

Reading Standards for Literacy in Science and Technological Subjects: Grades 6–8: Distinguish among facts, reasoned judgment based on research findings, and speculation in a text. (p. 62)

Is this a reasonable educational goal or more abstract word play? Again, the problem of relationships is of paramount importance. In order to distinguish between what research evidence supports and what is yet another example of scientism, where the scientist is extrapolating beyond what is supported by her/his own research findings as well as those of others, wouldn't the student need to be knowledgeable about what distinguishes science from scientism? How many students, for example, are able to distinguish between the nature and role of a gene in the process of natural selection, and the nature and role of a meme in the process of cultural evolution? Which is based on scientific evidence, and which is based on the speculation and the hubris of scientists? Are the claims that E. O. Wilson makes in his book *Consilience: The Unity of Knowledge* (1998) to be understood as based on research, or yet another example of speculation? He claims, for example, that there were no important advancements in knowledge until the emergence of Western science, that all cultures should abandon their religious traditions in order to base their beliefs and values on Darwin's theory of natural selection, and that scientists should determine the beliefs and values people should live by. Given Wilson's prestige as a scientist, how many students would be able to recognize when science becomes scientism? Would students be able to recognize that the scientist's recent support of intelligence testing and eugenics are examples of scientism? Would they be able to recognize that the current claims by computer scientists such as Ray Kurzweil and Hans Moravec that we are entering the post-biological phase of evolution, with super-intelligent computers replacing human intelligence, are also examples of scientism? And how many students, and their teachers for that matter, would be able to recognize the scientism in the thinking of Carl Sagan, who wrote *The Demon-Haunted World: Science as a Candle in the Dark* (1997), or the ideas of Francis Crick (co-winner of the Nobel Prize for his research on the structure of DNA), who claimed in *The Astonishing Hypothesis: The Scientific Search for the Soul* (1995) that scientists will shortly be able to explain the genetic basis of becoming a musician?

The number of scientists who are making claims about cultural developments for which there is no scientific evidence, which instead represent what Wendell Berry has referred to as their imperialistic intent to exclude all forms of knowledge other than their own (2000), is a potential source of danger. As we can see in how Ray Kurzweil and other computer scientists are predicting that robots will displace the need for human workers, and in how other computer scientists are hard at work in bringing this about, there are few scientists considering the social chaos that is likely to follow. Scientism is based on the same myth of progress that drives the thinking of market liberals and libertarians who assume that there are no limits to what capitalism can achieve. The questions become, what are the limits of scientific knowledge, and what are examples of scientism becoming the basis of unjust social policies? Unfortunately, these questions are not raised by the experts who are responsible for the CCS reforms.

Again, the problem of evaluating the student's performance in terms of meeting this standard becomes of paramount importance. This is yet another example of how learning standards are treated as isolated skills or fragments of knowledge totally divorced from the living cultural/linguistic contexts that are inescapable aspects of daily life. To enter the world of learning standards is to enter the world of abstract thinking that has been one of the dominant legacies of elite thinkers in the West. Yet, the world of abstractions where the flow of life-shaping processes can be reduced to what can be digitized is what the students are really being socialized to accept as their taken-for-granted world.

Literacy in Science and Technology Subjects, Grades 11–12: By the end of grade 12, read and comprehend science/technology texts in the grades 11–CCR text complexity band independently and proficiently. (p. 62)

Leaving aside the nearly incomprehensible manner in which this standard is written, the question of context and thus of relationships becomes important. Suppose students were to encounter a scientific report written by scientists paid by a corporation such as Exxon Mobil to challenge the evidence of global climate change and the evidence that human practices are contributing to it. The motive behind this effort to pay scientists to spread disinformation and to sow the seeds of doubt was to change public opinion, thus allowing the corporations to continue their profit-making, ecologically destructive practices.

Would students need background knowledge of the ideology that corporations rely upon to justify placing profits above slowing the rate of environmental degradation, and the catastrophic social consequences that will follow? What about the texts of computer scientists such as Ray Kurzweil, who predicts that computers will evolve to the point where they are capable of religious experiences, and Gregory Stock, who predicts that all of the world's culture will shortly be transformed into a "Global Superorganism"? Again, the question arises: How can the student's progress in achieving this standard of scientific and technological literacy be measured? How many of the programmers who are to create the computerized tests are able to explain the cultural non-neutrality of technology? How many of them know about the different traditions of ethno-science that have led to many discoveries from which Western scientists are just beginning to learn? An important characteristic of the entire CCS debate is that the depth of knowledge of the teachers and people who will be preparing the tests has not come up for discussion. As I shall explain in the next chapter, there is no evidence that what are regarded as the critically important learning standards that will enable students to succeed in world of work and in higher education take account of the multiple crises that we now face. Nor do they take account of the ideologies and cultural traditions that are partly the root of these crises.

What Is Worthwhile in the Common Core Standards

It is important, however, not to leave the impression that all the standards are too simplistic and ideologically naïve. A number of the standards emphasize the importance of giving close attention to what is actually written in a text or said in an oral presentation. Also, the emphasis on learning to give attention to the vocabulary of different domains of inquiry, and on being able to recognize and use proverbs as well as common idioms, is also important. A number of the standards recommended for different grade levels include such common-sense recommendations as "choose words and phrases to convey ideas precisely" (Grade 4, p. 29). And perhaps most important of all, learn how to "expand, combine, and reduce sentences for meaning, reader/listener interest, and style" (Grade 5, p. 29).

These are common-sense recommendations. But do they and the few other recommendations that are on the same level of common sense justify the hundreds of millions of dollars that are to be spent on implementing the CCS reforms in the 45 states that have adopted all or part of them? The question also needs to be asked again about the real motives of the business leaders, politicians, and educational leaders who are promoting these reforms so vigorously. Exxon Mobil Corporation has produced a slick television commercial promoting how the CCS reforms will lead to success in careers and in universities, and thus to a more competitive workforce in the global economy. The question that should be asked is whether the support for the CCS would disappear if the standards included what students should learn in order to exercise communicative competence in guiding the transition to a world of ecologically sustainable cultures.

Chapter 4

21st-Century Challenges: Will Students Be Prepared to Address Them?

When reading the justifications for the Common Core Curriculum Standards I am reminded of how the advertising industry operates on the premise that if promises are put into print or promoted through visual images most people will accept them as accurate representations of reality. The creators and supporters of the CCS reforms claim that they will lead to more "rigorous" thinkers, that they require "mastering a few things more deeply," that they prepare students for "college, workforce training, and life in a technological society." The most questionable claim is that "the Standards also lay out a vision of what it means to be a literate person in the 21st century" (2010, p. 3). These claims, and the educational consultants working on the standards that will supposedly achieve these high-sounding educational goals, have won the support of an impressive list of supporters. They range from the National Governor's Association for Best Practices to the Council of Chief State School Officers. Even the heads of major corporations ranging from the Boeing Company to Dell, GlaxoSmithKline, IBM, Intel Corporation, Microsoft, and State Farm Insurance (to name just of a few of the 70) signed letters of support.

The question not being raised about how the CCS reforms will propel the country forward in the global economic race, and prepare a literate citizenry that does not repeat the environmental mistakes of the last centuries, is whether any of these blue-ribbon corporations and political groups understand the major issues that have gone unaddressed over the last centuries—and have now reached a crisis point.

Becoming a governor, the head of a major corporation, a leading figure in a state office of education, or an expert in identifying learning objectives that can be tested should require becoming literate—that is, being able to read the cultural patterns that are at the root of the ecological crisis. But this is not the case. Indeed, it is important to note that the ecological crisis is not even mentioned in any of the CCS promotional literature, or in the list of learning standards that are to be tested. It is also important to note that the criticisms of the CCS reforms, mostly by so-called conservative political groups, do not mention the ecological crisis that has moved from the realm of abstract newspaper accounts to actually impacting people's lives in the form of extreme weather patterns, toxins that are affecting people's health and the diversity of species, droughts that are impacting agriculture and the availability of potable water, and the warming and increased acidification of the world's oceans that are reducing essential sources of protein.

In short, what is not being mentioned by either supporters or critics of the CCS reforms is now the dominant reality of the 21st century that educational and political reformers will be forced to address. As we can see in the drought-stricken parts of the country, in the forest fires now occurring on a biblical scale, and in the collapse of fisheries off the shores of the New England states and the Eastern Canadian provinces, radical changes in the behavior of natural ecological systems are forcing policy changes at the local and national levels that would not even have been considered a few decades ago.

There are two major points that need to be made about the educational backgrounds of both the supporters and critics of the CCS reforms. The first is that the major problems of this century are being intensified by the ideology that underlies the economic agenda of corporate America, whose CEOs came out in strong support of the CCS reforms. That is, the so-called conservative ideology represents the continued dominance of the paradigm that gave us the first industrial revolution, which is now in its digital phase of globalization.

As few university graduates are aware of the reflective and community-centered conservatism of Edmund Burke, and that important environmental writers such as Wendell Berry and Vandana Shiva are representative of the kind of conservatism that should be taken seriously in the 21st century, they have adopted the historical misconceptions that have led to today's faux conservatism. It is a mix of social Darwinist and market liberal assumptions about the progressive nature of competition, the autonomous individual who can survive only by relying on consumerism, the need to increase profits by replacing workers with computer-driven machines, and the imperative of relying upon digital technologies that fosters the cultural amnesia eroding the traditions of community self-sufficiency and mutual support.

The second problem is that aside from scientists who are studying the behavior of natural systems and issuing warnings about the cataclysmic changes that will overtake the world in the next few decades, there are few faculty outside the sciences who have overcome the silences that characterized their own years of graduate studies. Scientists, for example, are now predicting that by midcentury, as much of 40% of the world's population will not have access to adequate supplies of water. They are also reporting on how fish populations are migrating in response to changes in the temperature of the world's oceans. As the migration patterns are toward the ocean's cooler temperatures, large numbers of people living in regions where ocean temperatures are rising will be without their traditional source of food. Scientific studies of melting glaciers, the disappearance of species, and other ecological changes that bring into question the future of humankind stand in sharp contrast to what the majority of faculty in the social sciences and humanities, as well as in the professional schools, consider as being of scholarly importance.

There are, of course, important contributions being made in each of the disciplines and in professional schools. But even these achievements make only a limited contribution that falls far short of providing students and those in policy-making positions the conceptual background necessary for understanding the historical and current cultural forces now threatening the world's diversity of cultural and natural ecologies. There are now one or two faculty members in most disciplines who are breaking the silence about the cultural roots of the ecological crisis by using their disciplinary perspectives to focus on environmental issues. The emerging field of ecocriticism is an example of a more fundamental change in the focus of a discipline. In other departments such as history, anthropology, philosophy, sociology, and political science, an environmental focus is limited to a few of the faculty. Business schools have learned to straddle the fence between environmentalism and their traditional commitment to promoting the ethos of capitalism by popularizing the words *sustainability* and *green,* which corporations now use to expand further their market appeal. Faculty in colleges of education, like other faculties across campus, continue to promote the deep cultural assumptions about the culturally transformative nature of critical thinking (as if the combination of capitalism and digital technologies were not doing enough in this area), a human-centered view of the natural world, and the need to become computer literate. Like most of their colleagues in other departments across campus, they continue to perpetuate the silences and misconceptions of their graduate school mentors who were largely unaware of the ecological crisis and that its roots were partly in the technologies and deep cultural assumptions that can be traced back to earlier Western philosophers and social theorists.

The important point is that the faculty who are beginning to address the cultural/environmental connections are still in the minority, and that the men and women who have created the CCS, and the corporate and political leaders who are supporting them, were educated in universities when Earth Day celebrations represented one of the few times in their university experience that they encountered any mention that humans were changing the environment in unsustainable ways. Even if any of the key people who have shaped the agenda of the CCS had encountered a professor who introduced them to environmental issues and key thinkers such as Aldo Leopold and Rachel Carson, what they learned in their other courses would have diverted their attention back to the fragmented understandings that result from taking a wide mix of courses that have little in common except for the widely held misconceptions and deep cultural assumptions circulating through the courses like a not yet recognized deadly virus. These misconceptions include the assumptions that underlie the idea of objective knowledge, that the printed word refers to real entities and rationally based ideas, and that there is a connection between the exercise of individual intelligence and freedom.

21st-Century Issues That Are Ignored by CCS Reforms

The claim that teachers will still exert control over the areas of the curriculum not dictated by CCS standards, and subjected to the regime of tests, does not really address whether they will prepare students for "life in a technology society" and provide a "vision of what it means to be a literate person in the 21st century." Given what most classroom teachers will not have encountered in their professional studies, as well as in the courses taken in other disciplines, these claims are just so much empty rhetoric. The following are just of few of the dominant issues that will affect the lives of everyone but will not be introduced by classroom teachers. And in not acquiring even an elementary understanding of these issues, the students, like today's Tea Party and libertarian extremists, will be easily swayed by the demagogues in the pay of powerful elites trying to hold onto their privileges and wealth.

To determine whether my observations are accurate about what will be missing in the part of the curriculum that teachers control, it is necessary to consider what is missing in their own professional studies and academic courses. Like so many past reforms such as No Child Left Behind and earlier ones based on either the need to meet the challenges of foreign competition or more liberating goals that emphasized the student's right to decide what she/he wants to learn. Years of trying to get university faculty to focus on the deep cultural patterns they take for granted, and that are continually reinforced through the multiple communication pathways in their disciplines and in everyday life, has led to viewing the prospects of achieving the necessary reforms as exceedingly problematic. Nevertheless, the effort needs to be made because there may be others who may be more effective in carrying the reform project forward in coming years.

It is important to keep in mind the differences between how I represent the characteristics of language, technology, science, and so forth, and how they are represented in the curriculum of the CCS. It is also important to recognize that if students lack a basic knowledge of these cultural patterns and how they interact with the patterns within the interdependent cultural and natural ecologies, they will be unable to exercise the communicative competence required for participating in the democratic process that must be renewed at the local level, where issues of community self-sufficiency will become increasingly important.

Characteristics of Languaging Processes Critical to Addressing 21st-Century Problems

Brief review: This is a world of interacting relationships within the micro and macro cultural and natural systems. That is, there is nothing in this world that is totally autonomous, or free of being influenced by the information flowing though the multiple pathways connecting and altering the different systems that constitute our intersubjective and external worlds. All forms and levels of life depend upon multiple semiotic systems and pathways for communicating the difference which makes a difference. Moving a rook on the chess board sends a message to which the other player responds; a change in the tone of voice sends a message, just as a change in the temperature will prompt changes in the behavior of a plant or animal. Even silence sends a message that leads (depending upon the culture) to different responses—which lead to further messages or differences which make a difference. Contrary to how the world is misrepresented in the learning standards of the CCS reforms, where (to cite one example) "students employ technology thoughtfully..." as though technology is an inert object, the reality is that the technology (depending upon its nature) will actually amplify or reduce different dimensions of the student's experiences. Determining the "meaning of words and phrases," to cite another example, will depend upon the history of the linguistic ecology that has become part of the student's taken-for-granted world of understanding, and so forth. And the taken-for-granted understanding of the "experts" who design the tests will make a difference—but not in the direction of providing objective test results. Nothing exists outside of the interdependent cultural and natural ecologies, and because the cultural and natural ecologies have a history of earlier influences that will influence their current and future behaviors they represent an emergent world.

Basic Characteristics of Language that Students Should Understand

Misunderstandings about the nature and role of words in the process of communication:

Most teachers and professors, and their students, who now control the media and other forms of messaging, reinforce the mistaken idea of communication as a process of **sending** and **receiving** ideas, objective information, data, and visual images. This misconception is called the conduit view of language, and it has the effect of reinforcing another misconception that is hugely important: namely, that words refer to real entities, experiences, and ideas. The conduit view of language has the effect of hiding that words have a history, and that they are metaphors whose meanings were framed by the analogs settled upon in earlier times. Examples include thinking of a woman as having limited attributes, which reflected the prejudices of the people who chose the attributes as examples of what a woman is like. Thinking of the environment as wild and a source of danger, then later as an economic resource to be exploited, and now as an ecosystem, are examples of how over time the meaning of a word is given a new meaning that reflects more current understandings. Metaphors such as *progress, data, individualism,* and *intelligence* also have a history, and their current meanings were in part influenced by the cultural assumptions taken for granted by those who settled upon the analog that framed the word's current meaning.

Implications teachers should be able to bring to the attention of students:

Words (as metaphors) have different meanings in different cultures and at different times within the students' own culture.

Words (metaphors) whose meanings were formed by the choice of analogs settled upon in the distant past, or within a different culture, carry forward the misconceptions and silences of that earlier era. This is one of the reasons that students should be aware of the need to question their taken-for-granted patterns of thinking, as they may be based on earlier forms of intelligence encoded in the choice of analogies that framed the meaning of words. Gender-biased patterns of thinking, as well as thinking that this is a human-centered world, are examples of relying upon earlier forms of cultural intelligence—or what can be called the linguistic colonization of the present by the past.

The spoken and written uses of words are only part of the process of communication within and between cultural and natural ecologies. There are other forms of communication that can be understood more generally as signs—such as how chemicals interact with each other, how the code in the DNA leads to the sequential development of organisms, how animals and humans use body language to communicate about relationships, and so forth. Visual images, the architecture of buildings, and the organization of social space are also complex semiotic systems that communicate messages, just as droughts and extreme weather patterns are also signs that communicate in ways that set off complex messages in and through the ecological patterns that connect.

Most of the information or messages being communicated through the information pathways that connect cultural and natural ecologies will be ignored if the languaging processes are understood as involving only the spoken and written word, and if it is understood that words have a specific meaning that can be determined by the student.

Students should discuss whether the use of English nouns undermines awareness of relationships and contexts, and whether verbs are more suited to communicating about the emergent world of relationships. These same questions can be asked about the role of nouns in other languages, and whether some languages rely more or less upon the use of verbs.

Print is a technology that has many important uses but also fundamental limitations.

Print enables information and ideas to be shared widely, thus enabling people to gain from the insights of others. It also provides sources of pleasure when encountering the works of gifted writers, and is essential to legal documents. It also provides alternatives to oral narratives in recording historical events, and in fostering critical analysis.

As digital technologies become the dominant source of cultural storage and communication, print will increasingly displace reliance upon oral communication, and this in turn reduces intergenerational mentoring and the sharing of intergenerational knowledge and skills that mostly have a smaller adverse ecological impact.

While print has many important, indeed indispensable uses, it also has many limitations—none of which, like so many characteristics of language, are mentioned in the CCS. Print can represent only a fixed or snapshot view of a dynamic world. That is, it cannot provide an accurate account of the ecology of local contexts, where much of the communication depends upon tacit and taken-for-granted understandings. It can provide only a surface knowledge, and that which it gives an account of is frozen until a new printed account appears.

Print, as pointed out elsewhere, relies upon the conduit view (sender/receiver) of language, and also hides awareness that words are metaphors whose meanings were framed by the choice of analogs in the distant past, or by another culture. That is, print hides the fact that words (as metaphors) have a history, and that the analogs that frame the meaning of words change over time.

Print is inherently ethnocentric, as it reinforces a different form of consciousness than is found in oral cultures where consciousness relies more on memory, the senses, tacit awareness of local contexts, and that life is an emergent, dynamic, and interdependent process.

Print fosters abstract thinking where what is in print becomes accepted as an accurate account of reality (such as in an ideology) that no longer is related to the emergent world of cultural and natural ecologies. Thus, print leads to thinking in terms of universal truths and moral absolutes. As the abstract world is taken to be what is real, it is assumed to be an accurate representation of how everyday reality is to be organized. The abstract world constituted in print-based theories and accounts also marginalizes awareness of other cultural ecologies. An example of how print creates a world of universal truths can be seen in how Adam Smith's explanation of free markets, which ignored how markets operate in different cultures and in everyday life, has become a universal truth now being imposed on the entire world.

Print tends to hide the ecology of cultural and natural forces that have influenced the thinking of the writer. The widely promoted myth of objective thinking, which is based on a number of misconceptions, further hides that what is encoded in print, if traced back to its human origins, is an interpretation that has taken on the appearance of an objective account.

Characteristics of the Increasingly Technologically Mediated World We Live In

Brief Review: A careful reading of the standards that are to be met in the area of technical subjects are really focused on learning to read technical manuals, and on being able to follow the progression of steps required in dealing with the operation of different technologies. The claim of "Literacy in … Technical Subjects" and the equally hyperbolic promise that students will learn to "use technology and digital media strategically and capably" has served an important purpose in gaining the support of corporate executives, politicians, and educators whose don't-rock-the-boat mentality enabled them to reach the top of the decision-making pyramid. Unfortunately, it hides a basic reality that needs to be recognized if the world's cultures are to avoid the colonizing nature of the digital/corporate/militarized revolution now sweeping the world. What the CCS reforms will not enable students to understand are the cultural transforming and environmentally destructive characteristics of technologies. Instead, students will continue to think of technology as culturally neutral, and that the purposes of its users determine whether it is a constructive or destructive force. Most students, like the heads of corporations, politicians, and educational leaders, will continue to associate technological innovations with social progress—and will not ask about the diverse forms of cultural knowledge and mutual support systems that are displaced by new technologies. The statement in the CCS promotional statement that "students employ technology thoughtfully…" is again profoundly misleading. The following are some of the characteristics of all technologies, and especially digital technologies. Classroom teachers should be able to introduce students to the cultural and environmental non-neutrality of different technologies.

Basic characteristics of technology that students should be able to understand:

How to overcome the misconceptions that equate technological innovations with progress and claim that they are culturally neutral:

As Don Idhe has pointed out (1979), technologies have different amplification and reduction characteristics that alter human experience and impact the environment in different ways. Print amplifies the ideas of objectivity and factualness while reducing awareness of human interpretation. The pre–digital phone amplified voice over distance while reducing awareness of the nonverbal patterns of communication that are often the source of the more honest aspect of communication. Robots amplify speed, efficiency, and profits but reduce the importance of craft knowledge and skill. Robots also amplify the central control of the production process. Craft knowledge and skills involve a different ecology of amplification and reduction characteristics. Examining changes in both human experience and the environment can lead to becoming aware of what Monsanto's Roundup herbicide amplifies and reduces, or what is amplified and reduced when learning is mediated by the use of computers.

The amplification and reduction characteristics of technologies that range from a cookbook to the Internet suggest the importance of thinking of a technology not as an isolated entity, but instead as part of the larger and multilayered cultural and natural ecologies within which it is used. Examining the amplification and reduction characteristics, which will change in different ecological systems, is another way of understanding what Bateson meant by saying the difference which makes a difference is the basic unit of information (here the metaphor of information needs to be understood in the broadest and most encompassing sense).

The characteristics of the cultural non-neutrality of digital technologies that students are to thoughtfully employ:

Among the most important losses associated with digital technologies are the forms of knowledge, experience, tacit knowledge, and taken-for-granted cultural patterns that cannot be digitized. In short, what is not explicitly known cannot be digitized, just as it will not be put into print. Oral communication, which involves many different information pathways that are largely tacit and dependent upon shared taken-for-granted understandings, often occur below the level of explicit awareness. Digitizing specific moments or events in oral communication reproduces the limitations of print that provide only a surface understanding of a more complex ecology. Video streaming of oral communication or performances reproduces more of its complexity, but again, only at the surface level. The intersubjective world of the participants is still inaccessible and thus cannot be digitized.

What digital technologies are able to make available to their users is, in the main, what appears on the screen in the form of print, images, graphs, and flowcharts. As in the case with print, the assumptions and thus the interpretative frameworks of those who write the programs that organize what has been digitized are hidden. Again, the information pathways that sustain the cultural and natural ecologies from which the data is derived are marginalized. While digital technologies can more accurately represent the interdependent patterns within both cultural and environmental ecologies, they still fall short—which then opens the door to generalizations that do not take into account the specific cultural and environmental ecologies.

Digital technologies, driven by both the cultural assumptions of computer scientists and by the need of corporate capitalism for an endless stream of technological innovations, amplify that this is a world of continual change—which is also represented as progress. These assumptions, which include the idea that we are autonomous individuals, reduce awareness of what needs to be conserved—such as the intergenerational achievements in the areas of social justice, knowledge, and skills that contribute to the self-sufficiency of the community and that have a smaller adverse ecological footprint.

The amplification and reduction characteristics of digital technologies—ranging from computers to iPads, cell phones, social networks, and so forth—are being found to alter the nature of consciousness itself. The reduction characteristics include the loss of long-term intergenerationally informed memory, a shortened attention span, and a proclivity to be looking at a screen rather than establishing and maintaining eye contact with others—and engaging in reciprocal nonverbal patterns of communication. Similar to the way focusing on print reduces awareness of the multiple forms of information being communicated within and between the cultural and environmental ecologies within which the person is situated, digital-based storage and communication reduces awareness of what various ecological systems are communicating about the changes they are undergoing.

While digital technologies that provide a better understanding of how changes in the patterns connect between ecological systems are a definite improvement over the more static way in which print represents interactive relationships, they nevertheless reduce the individual's ability to exercise the ecological intelligence that is dependent upon recognizing and responding to the actual flow and range of information between and within the different ecologies that are experientially present. Reading a manual describing how to pass another car safely is less useful than being able to observe and calculate the effects of different weather conditions, the behavior of other drivers, and the necessary space and speed needed for passing safely. The latter is an example of exercising an individually centered ecological intelligence.

The increasing reliance upon digital technologies, in spite of the return of small-scale production promised by the introduction of the new additive technologies such as the 3D printer, amplifies the centralization of decision making, which in turn reduces the worker's role in local decision making. Indeed, the extent to which production and distribution of goods, as well as manipulation of the consumer's consciousness, is now digitally controlled reduces the need for workers who perform routine tasks.

The digital revolution is also a revolution in the area of civil liberties. That is, civil liberties such as privacy in communication and in activities and relationships with others, and increasingly in thought itself, have largely given way to a condition of being under constant surveillance that is more appropriate to a police state. This can be understood as amplifying the potential and actual exercise of the government's police powers over the activities of citizens. With the introduction of Big Data technologies into workplaces, data can be obtained on the workers' on-task behaviors. This data enables management to use its reward and penalty systems to re-engineer the workers' movements in order to increase efficiency and profits. Again, the amplification is in producing more profits while reducing the quality of the work experience.

The myth that represents technologies as a tool whose meaning is determined by the person or group using it is unlikely to be challenged by the people who are identifying the learning standards in the CCS. This is primarily because few of these curriculum and testing experts will have encountered in their university experience an in-depth examination of the amplification and reduction characteristics of different technologies as they are introduced into different cultural and natural ecologies. What they failed to learn is what is being passed on to the next generation, but hidden by printed words that hide what is missing.

Learning to "Read" the Cultural Patterns that Will Enable People to Live Less Consumer-Dependent and Environmentally Destructive Lives

Background Review: There is nothing in the Common Core Standards that addresses the ecological/cultural crises. The primary reason for this is that the business leaders, politicians, and educational experts who are promoting the CCS reforms have shown no awareness that we are on the cusp of having to make radical changes in our consumer-dependent lifestyles. As discussed earlier, natural systems are already introducing fundamental changes that will quickly move from appearing to be beneficial in food production to forcing major disruptions as droughts spread and as disease-carrying organisms migrate northward to warmer climates.

While some of the CEOs of major corporations are aware of environmental issues, they have not altered their drive to increase production, market share, and profits—all of which contribute to degrading further the natural systems we depend upon. The politicians as well as the educational experts

have shown no awareness that we are on the cusp of life-altering changes due to environmental changes and to the digital revolution that is contributing to loss of employment for millions of people around the world. Poverty is both deepening and spreading, while the super-wealthy continue to exert greater control over the political process at both the federal and state levels.

The most viable alternatives to the cultural forces that are further degrading the natural systems and the prospects of employment for increasing numbers of people in the United States and abroad are the diversity of cultural commons that already exist in the world's cultures, and thus in every community. As will be explained, the knowledge the student most needs in the 21st century is how to "read" (recognize and interpret) the patterns of intergenerational knowledge and mutual support that reduce dependence upon a money economy that is becoming increasingly unreliable for many people. In short, students need to be able to recognize the traditions of the cultural commons that support community self-sufficiency and local democracy—as well as the traditions of prejudice and environmental abuse. They also need to be able to "read" the many ways in which technological and ideological forces are enclosing what remains of the cultural commons—that is, integrating them into the market system that increases dependence upon a money economy that is being manipulated in ways that keep large segments of the population living in poverty. The CCS reforms, particularly in the area of literacy in technical subjects, will leave students with such a naïve understanding of technology that they will be unable to recognize how they may be contributing to further displacement of workers and skills.

Basic characteristics of the cultural commons and the forces of enclosure:

Learning to "read"—that is, to recognize—the gift economy of the cultural commons.

The dominant characteristics include knowledge, skills, and mentoring patterns passed on through face-to-face communication over many generations. This intergenerational heritage includes life-sustaining practices in the areas of food, healing, ceremonies, narratives, creative arts, craft skills and knowledge, games, knowledge of the behavior of local ecosystems, patterns of moral reciprocity, local decision making, and language itself.

The intergenerational knowledge and skills will vary from community to community, from culture to culture, and from bioregion to bioregion.

The cultural commons represent a gift economy, as knowledge and skills are freely shared with the understanding that they will not be monetized by future generations. All communities, even those that have traditions of oppressing others and degrading the local environment, have traditions of mutual support in all of the above areas. An ethnography of the cultural commons in any community will reveal that the cultural commons is not an abstract idea, but a living reality that represents alternatives to the many forms of dependency required by a market economy where profits are more important than the welfare of people and the environment.

Cultural commons practices reduce dependence upon consumerism as they promote the development of personal skills and mentoring relationships—and a sense of community identity.

Since making useful things and sharing knowledge and skills reduce dependence upon the industrial system of production and consumption, the cultural commons have a smaller adverse ecological footprint. This can be recognized by comparing the toxic chemicals used in the industrial process with the few used in the local production of food and in other cultural commons activities.

As cultural commons activities are not driven by the requirements of industrial and profit-driven schedules, people can engage in less stressful activities, and in being less dependent upon a money economy, there is less of the stress that comes with the fear of being replaced by a machine or a younger person who will work for less. What needs to be investigated is whether people engaged in cultural commons activities have fewer health problems that can be traced to the toxic and stressful environments of the market-driven workplaces.

Learning to "read" (recognize and interpret) the cultural forces that are enclosing the cultural commons by integrating what is freely shared into the market system that requires participation in a money economy now governed by market liberal/libertarian principles.

Recognizing the experiential differences between a cultural commons and market-based experience—in the preparation and sharing of food, in participating in a creative arts activity, in a game, in the exercise of a skill in producing something useful, in being in the natural environment. Which leads to discovering an interest, developing a talent, in being part of a community.

Being able to recognize how different technologies enclose and thus lead to the monetization of the cultural commons. For example, does print and now, digital-based storage and communication undermine the face-to-face communication that is part of a mentoring relationship? Does the increased reliance upon the Internet contribute to the idea that the knowledge and skills of older family and community members are outdated and largely irrelevant in a culture undergoing a radical transformation? What skills and knowledge traditions may be lost as the current generation becomes fully engaged in the cyber culture? Can what is being lost, such as privacy, civil liberties, knowledge of how to preserve food, how to reduce the need for toxic chemicals, and how to increase the life chances of insect pollinators so essential to the availability of food, and so forth, be recovered if there is no memory, and if the dominant ethos is to look to the next technological innovation?

How do the silences about the nature and importance of the cultural commons on the part of classroom teachers, professors, media gatekeepers, and promoters of corporate products impact the ability of the world's cultural commons to survive as alternatives to the consumer-dependent lifestyle that is undermining the viability of natural systems? Should this question be raised in the upper grades by teachers who make decisions about the curriculum? Do teachers have the background knowledge to even understand the importance of the question?

Should students be able to recognize the deep cultural assumptions inherited from earlier generations that continue to provide conceptual direction and moral legitimacy to the industrial culture that, in its digital phase of globalization, is undermining the cultural commons that represent all that stands between a subsistence existence and death for hundreds of millions of people? Will the CCS reforms contribute to being able to "read" the cultural patterns contributing to the spread of poverty?

The cultural commons are being enclosed through the loss of historically informed understandings of language, such as the way the word *wisdom* is now being replaced by data and information. The current Orwellian use of our political language is yet another example. For example, the word *conservative,* which requires critical thinking that identifies what needs to be changed as well as what needs to be intergenerationally renewed, now is used to justify the economic and social agenda of market liberals and libertarians who adhere to the survival of the fittest ethos of social Darwinism. Where in the CCS reforms is there the suggestion that students should understand that our political language has a history? It would seem that becoming "a literate person in the 21st century" would require giving attention to what they need to know in order not to be swept along by ideologies that have their roots in earlier centuries when there was no understanding of environmental limits, and attention to the other cultural ways of knowing that are less individually and competitively centered. If this is important, then the committees that created the CCS reforms should have addressed the shortcomings in the education that most teachers received.

The CCS reforms also should have addressed the teachers' pedagogical responsibilities as mediators in helping students recognize the difference in their cultural commons and market-oriented experiences. The role of mediator involves helping to make explicit differences in experiences, rather than promoting an ideology that is based on the deep cultural assumptions that underlie the current myths about progress and individualism. It also requires awareness of cultural patterns that others take for granted, and the ability to provide a historical understanding of the cultural patterns.

"Life in a culture that produces more than eighty two thousand industrial chemicals that go into the manufacture of more than ten million products will affect the prospects of today's students who will only encounter the silences in the CCS reforms and in the curricular decisions by teachers"

Background Review: The linguistic colonization of the present by past ways of thinking, the constant interactions with technologies, and the enclosure of the cultural commons go largely unnoticed. The impact of the thousands of toxic chemicals on the development of children and on the health of adults has reached a crisis point. The medical costs are staggering, and the number of wasted lives is immense—with some scientists even concerned that the impact of pollutants on the brain is leading to a reduction in intelligence. The moral code governing the experimental use of synthetic chemicals is based on the assumption that they can be used until evidence suggests they need to be controlled. As many of the chemicals used in the manufacturing process are released over many years, it is difficult to determine which are safe until it is too late.

As learning about this aspect of the scientifically based industrial/ consumer-dependent culture is largely ignored until the latter stages of graduate school, most Americans continue to view the chemical assault on themselves and on the natural systems through the rose-colored glasses of progress. And those who are sounding the warnings about the dangers of toxin-saturated lives meet the fierce and well funded resistance of the chemical industry. So the question is: Should the CCS reforms in "Literacy in...Science and Technology Subjects" include the introduction of basic concepts related to the politics, economics, and health effects of this massive growth industry? Should teachers be required as part of their professional studies to take a course that covers the main features of the chemical industry that is largely out of control?

Following are a few suggestions for connecting different educational reform agendas that go beyond the dumbing down of the CCS reform proposals.

Learning to think of relationships rather than of discrete entities as a starting point in introducing students to the role of chemicals in life-forming and destructive processes.

Bateson's statement that a "difference which makes a difference" involves the initial exchange of information (signs, changes dictated by generic and chemical codes) between different micro and macro ecological systems. This is basic to understanding that this is a world of emerging relationships, rather than autonomous entities.

This basic understanding needs to be made explicit, rather than just assuming that students will understand. It can be introduced by asking students to think of examples within their own families of how different products are explained as having a damaging effect. What substances should not be given to animals, be ingested by humans, or used to nourish plants? What are the labels and warning that should be understood as signaling danger to health, and even death? This is, at a very elementary level, an introduction to relational thinking. But it does not enable students to understand the information pathways where chemicals interact with the development of organs and systems, such as the impact of certain toxic chemicals on the immune system or fertility.

Encouraging students to do an inventory of toxic chemicals in the house and the garage, as well as of the toxic chemicals in manufactured products brought into the home, will help to illustrate how pervasive toxic chemicals are in our environment—and how closely we are living with life-deforming toxins.

How language issues and the intergenerational knowledge carried forward in the cultural commons relate to the misuse of toxic chemicals.

The metaphors that reproduce earlier misconceptions that are not recognized because of their taken-for-granted status in our everyday thinking and communication are responsible for many of the misuses of toxic chemicals. Thinking of the ocean as too huge to be affected by dumping our waste and toxins is just now being recognized as a conceptual error. Middle-class values about what a yard should look like—green and free of "weeds"—has led to using fertilizers and herbicides to achieve a certain image dictated in part by shared visual metaphors. The word *weed* is thought of as a destructive and invasive plant that needs to be eradicated with toxins that move through a number of ecosystems before entering the local streams, rivers, and then the ocean. The analogic thinking that frames how we think of what a weed is like does not include an awareness of its nutritional qualities—which can be seen in how we deal with, for instance, dandelions. Cedar trees, to the European immigrants bent on clear-cutting the forests across America, were seen as lacking the qualities of the forests to which they were accustomed, and thus were treated as trash trees that need to be burned or left on the forest floor.

These examples suggest curricular possibilities that will introduce students to the connections between the metaphorical language they use to name their world and how this naming process influences their awareness of the plants and animals in their bioregion. What are the animals and insects that have been labeled as pests and as destructive, and what toxic chemicals are used to eradicate them? Do the metaphors that represent them as pests or as life-threatening take into account their role in the food webs they support? How we think about wolves is a good example of how cultural patterns of thinking, including misconceptions, have led to thinking of them only as destructive predators that needed to be poisoned or shot. The metaphor of "killer whale" that has been given to the orca whale (also known as the "blackfish" among the indigenous cultures of the Pacific Northwest) is an example of how a metaphor (and the comparative image it suggests) limits understanding the varied cultures, language systems, and social behavior of orcas.

What is often overlooked about the world's diversity of cultural commons that began with the first humans living in the savannahs of what we now call Africa is that these cultural commons included intergenerational knowledge of how to live within their bioregions without the use of today's synthetic chemicals. They often had shorter life spans, and even disappeared because of environmental mistakes. But many also developed, some to a high degree, their own ethnoscience of the properties of animals, plants, and soils that enabled them to solve problems. The pharmaceutical industries are now taking out patents on this wealth of enthnoscientific knowledge which has been shared over generations as part of their cultural commons related to healing practices, protecting crops, and so forth. There is still a legacy carried forward by ethnic and indigenous cultures of what they still regard as common-sense knowledge of how to deal with a variety of problems that most Americans turn to the chemical industry to solve.

Interviewing the older members of the community about traditional nature-based strategies for controlling pests in agriculture and in the care of animals reduces dependence upon the toxins produced by the chemical industry. These problems existed before the emergence of the chemical industry and the idea that there is a toxic chemical solution for every situation. Many of the traditional strategies were refined over generations without poisoning the environment.

"What it means to be a literate person in the 21st century" requires knowledge of the democratic processes, of the political belief systems that influence people's choices, and of what constitutes social and ecological justice. As the CCS reforms are based on the assumption that the curriculum that will address these aspects of the students' adult lives will be left to the judgment of the teacher, the question about whether teachers possess the depth of background knowledge necessary for introducing students to more than a romanticized understanding of the democratic process in today's society is not even raised.

Background Review: When we take into account the complexity of the democratic process—that the individual's ideas and values are largely influenced by the language that encodes the cultural assumptions that shaped earlier ways of thinking, by how the ecology of language of different ideologies frames

how problems are understood and limits alternative ways of knowing, and by how social and ecological justice are understood—it becomes clear that there are few current classroom teachers who are aware of any of the above cultural patterns. Their thinking is still rooted in the old paradigm that represents the individual as an autonomous entity (perhaps influenced by a constructive or destructive home environment); language as a conduit in a sender/receiver process of communication that hides that words, with only a few exceptions, are metaphors and that they have a history; and the democratic process as expressing one's own ideas and voting in an election. Before the curriculum was mediated through a digital technology, the dominant ethos held by many teachers was that students should be encouraged to construct their own knowledge. Computers, iPads, and other means of electronic communication have not altered this way of thinking of empowering individuals. This view of the educational process is summed up in the idea that teachers should promote critical inquiry and a transformational way of thinking. Seldom mentioned is that critical inquiry should also address what needs to be conserved and intergenerationally renewed. The teachers' lack of a historical perspective also influences their lack of understanding of the historical roots of current ideologies such as liberalism, libertarianism, conservatism, and socialism. Nor are they likely to understand that each of these political categories have mutated into environmental conservatism, religious conservatism, and market and social justice liberalism. Nor are they likely to recognize how Darwin's theory of evolution is being transformed into an ideology that supports the globalization of Western technologies and market systems.

Suggestions for introducing students to the political world that will shape their prospects of employment, allow them to live above the poverty level, and ensure the well-being of their progeny in an economic system that equates progress with exploiting natural systems:

1. *Reviewing the relational nature of existence—in one's intersubjective life, in culture, and in natural systems.*

As the misconception that this is a world of distinct entities, including thinking of ourselves as autonomous choice-making individuals, it is important to continually review the relational nature of experience and thus of existence. This would include engaging students in a discussion of how their identities are experienced as changing depending upon changes in the social and environmental context. Again, the importance of giving special attention to what is being communicated in and through these different relationships needs to be re-emphasized as the dominant culture continues to undermine awareness of the complex messages communicated through relationships—especially those that involve exploiting others, degrading the environment, and questioning what is represented as progress.

Another relationship that is widely ignored is how we are dependent upon an inherited vocabulary (system of root and iconic metaphors) that carries forward and reproduces many of the taken-for-granted ways of interpreting what is being communicated throughout the patterns that connect in this relational and interdependent world. Students need to be constantly reminded to consider the history of words, the earlier cultural assumptions that influenced the selection of analogs that frame their current meanings. This review should also include examples of how current thinking about political and environmental issues continues to be influenced by earlier misconceptions and silences. Reviewing the relational nature of all aspects of existence, from the molecular to the individual, and to the macro ecological systems, as well as how language carries forward earlier ways of thinking, reinforces an ecological understanding of the democratic process.

2. *Ideologies that support and undermine democracy.*

a. Ideologies are interpretative and moral frameworks that guide political decisions. They encode long-held cultural assumptions derived from the abstract theories of philosophers, as well as from past experiences. The vocabularies reproduce the cultural assumptions, and exclude the vocabularies based on different assumptions—including those of other cultures. The connections between past ways of thinking and the ideologies that guide current decisions that either support or undermine the democratic process can be seen by identifying the vocabularies of different expressions of conservatism (environmental, social justice, religious), liberalism (social justice, market liberalism), libertarianism, socialism, and social Darwinism. Identifying the vocabulary of what each of these ideologies excludes will also be useful in understanding how the language derived from the past limits what can be discussed in reaching decisions.

b. Identifying the cultural assumptions that underlie each of the ideological traditions should lead to questioning which assumptions take into account the interdependencies within both human and natural communities. In other words, which are based on an awareness of environmental limits, and which are based on abstract thinking that does not take account of different cultural ways of knowing? How does abstract thinking affect the process of finding common ground and compromising in ways that take account of the Other's interests?

3. *Democratic societies' history of genuine achievements in the areas of social and ecological justice, and also of injustices done to others and the environment.*

The social justice list of achievements includes the Constitution, the Bill of Rights, laws that protect workers (including children), protections of the legal system, and the recent gains made in providing equal access to education and the political process. At the same time, there are injustices that continue to be perpetuated, such as gender-based unequal pay and leadership possibilities, economic exploitation of different social groups, unequal educational facilities and opportunities, prejudices that influence the behavior of the police and the number of nonwhites held in prisons, exploitation of immigrant workers, an unfair tax system, ideologically oriented courts that enable corporations and the superrich to exert greater influence on state and federal policy making, and the development of digital technologies that are undermining traditional civil liberties. The ability to address any of these injustices is influenced in different ways by the deep cultural assumptions that underlie the different ideologies.

The gains in the area of ecological justice include the establishments of the national park system, as well as the creation of a number of state and federal agencies such as the Environmental Protection Agency, charged with overseeing public and corporate practices that are environmentally destructive. The continuation of ecological injustices include the excessive power of corporations to influence the legal process in ways that enable them to place profits ahead of contributing to an ecologically sustainable future, and the ability of corporations to promote a hyper-level of consumerism that does not take account of how it undermines the viability of natural systems.

4. The role of digital technologies in contributing to the unemployment of professional and skilled workers, in creating a surveillance culture, and in contributing to a form of consciousness that is more focused on digital screens than on social relationships and long-term memory.

a. The democracy that today's students will encounter as adults is undergoing radical changes due to the influence of digital technologies. Computer scientists and corporations are working to introduce more computer-driven machines that will increase profits while reducing the need for workers. Employment in the future will require knowledge of how to maintain and advance the development of digital technologies. Employment will also be available at low-paying jobs, as long as the cost of automation remains higher than the cost of paying below–poverty-level wages.

b. Students will need to be able to challenge this trend which is driven by the ideology of market liberalism. This will require a knowledge of and an ability to articulate the impact of the massive shift in wealth on the lives of the majority of people. They will also need to acquire the political skills necessary for organizing the vast number of people who are now being made redundant by robots. A knowledge of how to expand and revitalize the cultural commons should be part of a strategy of reducing dependence upon a money economy and hyper-consumerism. As corporations have demonstrated a long history of using violence to protect their interests, and now have access to surveillance technologies that can identify sources of dissent, it will be necessary to learn Gandhi's strategies of passive resistance that have been used in the recent civil rights and other social justice movements.

Concluding Remarks

The old adage about "cultural lag" contains an important element of truth. Why it occurs and what its impact is on addressing the most important issues facing this generation, which are the changes in natural systems caused by human behavior, is easy to explain. We have only to consider how being socialized into thinking within the language/thought patterns of one's culture makes it a nearly inevitable process. By recalling what was explained earlier about how today's meanings of words (metaphors) were largely framed by the choice of analogs settled upon in previous cultural eras that did not take account of today's social and ecojustice issues, we can more easily recognize why the patterns of thinking will, for most people, be out of step with today's most pressing issues. As cultural lag is a prominent characteristic of the thinking of the business leaders, politicians, and educational experts in testing and curriculum development, we have only to consider the silences and misconceptions encoded in the language most academics learned to take for granted in the last decades of the 20th century. Intelligence was understood as an attribute of the individual; language was not widely understood as an ecology of metaphorical constructions, but as referring to real and objective entities and relationships. Few understood the profound differences in consciousness and social relationships that separate oral from print-dominated cultures. The commons were understood only in terms of natural systems and how different cultures utilized them—with the nature and importance of the cultural commons existing only at the level of taken-for-granted practices that were below the surface of explicit and critical awareness.

As mentioned earlier, few university students in the last decade of the 20th century, which includes the people promoting the CCS reforms, were aware that technologies are a cultural transforming force, and that the reliance upon print and now, digitized visual images and voices, led to the loss of intergenerational knowledge that is less dependent upon a money economy and consumerism, and on the toxins now poisoning the environment and ourselves. That we live in a relational world (that is, in cultural and natural ecologies of constant information exchanges of which we need to be aware) was recognized only by a few academics who had read Bateson and the other people influenced by him, and who had learned from indigenous cultures their relational understandings of the world in which they lived. In short, the promoters of the CCS reforms were victims, like most people still in denial about the established consensus among scientists that human behavior and values are degrading the natural systems they depend upon. The cultural lag is evident even in the dominant ideological mix of market liberal and libertarian thinking that is based on the abstract 18th- and 19th-century theories of philosophers (and updated by theorists such as Ayn Rand and Milton Friedman), which did not take account of how markets function in both indigenous cultures and in the cultural commons of every community.

How to Address the Double Binds of Addressing 21st-Century Problems While Relying Upon the Thought Patterns of Earlier Centuries

Discouraging is too mild a term to describe my efforts to introduce faculty in teacher education, environmental studies, and such departments as English and philosophy, whose taken-for-granted patterns of thinking are still rooted in 20th-century assumptions and silences, to the ecology of language, the cultural commons, the difference between individual and ecological intelligence, and the noncultural neutrality of technologies. Silence has been their main response, though one faculty member at a Midwest university had the courage to ask which planet I was from. The shared indifference on the part of most faculty to the possibility that the ecological crisis might have implications for questioning the 20th-century conceptual orthodoxies they continue to pass on to the next generation of students, as well as their shared lack of the conceptual frameworks necessary for understanding what questions to ask or to explore, can be traced to the late 20th-century modernizing assumptions and explanatory frameworks they learned during their years of graduate studies.

The problem, in short, is that changes in the natural systems upon which we depend are occurring at a rate that now outpaces the rate at which taken-for-granted beliefs are changing. We have only to consider the centuries it took feminists to successfully challenge the metaphorical language that encoded the prejudices against women; we are witnessing the same slow pace in challenging the image of the environment as an endlessly exploitable resource. The awareness that digital technologies are contributing to the loss of intergenerational knowledge and skills continues to be buried under the weight of a blind faith in the supposed progressive nature of technological innovation—even when these technologies move us closer to the total surveillance culture of a police state.

There often are faculty members in different disciplines who are beginning to address environmental issues, and there is the possibility that they may focus on one or two of the issues discussed here. But the reality is that teacher education programs, including the courses taken in the departments of social sciences and the humanities, will not provide the basic conceptual background that classroom teachers will need in order to present students with a curriculum that compensates for the misconceptions surrounding the skills that will be tested. In the next chapter, I will focus on the most essential characteristics of key concepts and skills, and then suggest ways in which the students' largely taken-for-granted cultural patterns can be used to understand the issues that democratic citizens of the 21st century will need to address.

The challenge will be to explain how classroom teachers can learn to "read" the cultural patterns that previous and current generations of teachers and professors took for granted. What my efforts cannot bring about is the life-transforming awareness on the part of classroom teachers that there is an ecological crisis whose impacts are just beginning to be experienced in the extreme weather patterns, droughts, changes in ocean temperatures and declining fish populations, and loss of species. For classroom teachers, taking the ecological crisis seriously and recognizing that the pre-21st-century beliefs and values are major contributors is a personal decision—as will be the decision to make the effort to learn how to introduce students to the relational and interdependent natural and cultural ecologies within which they live.

Chapter 5

Classroom Practices That Avoid the Constraints of the Common Core Standards

Classroom teachers are caught in a three-way double bind. If the test scores on the content of the CCS are not high enough, school administrators will respond to outside political pressures by finding ways to penalize the teachers. Low test scores will further motivate corporations to make millions of dollars by providing new educational software and testing services. This will lead to replacing more classroom teachers with online courses that can also be tailored to reflect the beliefs and values of parents who currently homeschool their children or send them to the private schools that further erode support for public schools. The third double bind is that if teachers do not introduce students to the dominant environmental and cultural issues that must be addressed as the global environmental crisis deepens, and as the surveillance technologies take us further down the pathway leading to the police state which will protect the interests of corporations and the already superrich, rising levels of poverty and unemployment will lead to social chaos sparked by rage and a sense of hopelessness.

As the globalization of the digital revolution spreads, and the efforts of already rich and powerful individuals, as well as countries such as China, to obtain control of agricultural land in the Third World and other essential resources are stepped up, the shortage of basic resources will force people to engage in armed struggles for survival. The latest warning of what lies ahead is the recent report of the United Nations' Intergovernmental Panel on Climate Change (IPCC), which projects that the combination of increases in world population (an estimated 9.5 billion by 2050) and the 2% per decade decrease in available food-producing land will be matched by a 14% per decade increase in the global need for food. The report, *The Future of Employment: How Susceptible Are Jobs to Computerization?*, by two researchers at Oxford University, Carl Benedikt Frey and Michael A. Osborne, points to another crisis that lies immediately ahead. After undertaking a careful analysis of whether the characteristics of the 702 different jobs they examined lend themselves to being performed by software programs and robots,

Frey and Osborne concluded that over the next two decades, 47% of employment in the United States is at risk (Frey & Osborne, 2013). Unfortunately, if teachers continue to ignore these 21st-century issues as well as those identified in the previous chapter, their students will be even less prepared to understand how to bring about the changes in current cultural practices, including in ecologically destructive ideologies and technologies, which our future survival will require. Given that shopping malls are bursting with goods (many produced in China) and that media advertising is reaching new levels of aggressiveness, it might seem that these warnings are unfounded.

As the needed reforms in teacher education, as well as in the university curricula, generally are not likely to happen, we need to take a radical approach that breaks with the past. This will mean challenging the current efforts that rely increasingly upon educational software that continues to ignore that what students most need to make explicit and to critically examine are the cultural patterns they take for granted. The reference to the cultural patterns that are "taken for granted" is especially important for teachers to understand. Most schooling, whether in public schools or in universities, involves learning what has its roots in the printed word. That is, most of what is being taught and supposedly learned is the knowledge that has been made explicit, which is a requirement of print-based storage and communication. There are times when this has led to making explicit and examining what previously was taken for granted, such as sexist, racist, and homophobic attitudes and practices. Print also has been useful in expanding understanding beyond the individual's immediate experience, in regards to, for example, global environmental issues as well as the connections between power, wealth, and the politics of self-interest that are undermining the democratic process.

Taken-for-granted beliefs and practices are not always destructive, nor are they always politically, morally, and ecologically problematic. Cultural patterns and values that carry forward political wisdom refined over centuries of experience—such as the separation of church and state, privacy and other civil liberties, the practices of living light on the land, being mindful about what needs to be conserved that reduces human suffering while strengthening mutually supportive communities, and so forth—are examples of taken-for-granted practices and beliefs held by many individuals.

The point of making all taken-for-granted beliefs and practices subject to critical inquiry (*interrogated,* to use a word fashionable among certain educational theorists who have borrowed the metaphor that describes what precedes the act of physical torture) is that once what is taken for granted is made explicit and subjected to the critical inquiry of individuals who may bring with them what they often incorrectly assume to be their own ideological agendas, the wisdom of past achievements may be overturned. For example, making explicit the taken-for-granted practices and values that within many segments of society have led to treating women and other previously marginalized groups as equal and deserving of respect for their achievements may lead to repoliticizing what was previously settled as a social justice achievement. Questioning the taken-for-granted status about the need to separate church and state also leads to a step backward that creates an opening for extremist groups to promote their messianic agendas. The challenge facing teachers is to be able to identify which taken-for-granted cultural beliefs and practices support the well-being of communities and carry forward the social justice achievements of the past, while also recognizing the taken-for-granted patterns that benefit the few over the many, and that further degrade the sustainable characteristics of natural systems. Teachers face an especially difficult challenge in knowing which taken-for-granted practices of other cultures need to be made explicit and subjected to critical inquiry—and which to leave as part of a heritage that is not to be politicized. That most teachers lack a deep knowledge of the taken-for-granted world of other cultures makes their mediating role even more difficult.

The important point to take away from this all-too-brief discussion of taken-for-granted cultural practices and beliefs is that they are always present, regardless of what is being discussed, critically examined, and promoted as a reform. They are always present even in discussions of the uses of technologies and in the explanatory power of the sciences. Regardless of the subject, the teacher needs to ask the class (or in some instances, her/himself) what is being taken for granted that needs to be made explicit and what needs to treated not with further silence, but with an examination of the historical forces that led to what is now taken for granted.

Because of the power of taken-for-granted beliefs and values to resist new ways of thinking, it is important that the following suggestions for providing students the conceptual basis for exercising communicative competence in the world that will emerge over the next few decades be presented in simple, easily understood terms, along with examples derived from the students' own cultural worlds. Most important of all, the following explanations of the life-forming linguistic relationships need to be introduced over and over as students are engaged in learning other subjects—even those in the CCS curriculum. Without continual review, the taken-for-granted cultural assumptions that underlie the CCS, as well as curricular units introduced by teachers, will lead to overlooking that what is being learned is always based on hidden cultural assumptions. It's like what Clifford Geertz reported about his conversation with a peasant: When asked about what supports the world of sense experience, the peasant responded that it rested upon an elephant, and when asked what supports the elephant, the reply was that it rests on another elephant, and so on. Similarly, our world of daily experience is based on taken-for-granted assumptions, which in turn are supported by cultural assumptions rooted in the more distant past—all the way back.

As the taken-for-granted beliefs and practices are part of a larger ecology of taken-for-granted cultural patterns, they are often highly resistant to change. The following suggestions for challenging the taken-for-granted patterns of thinking and values are applicable to all areas of the curriculum. It is critically important that teachers are able to judge the level at which each of the following explanations and questions should be introduced. Each can be reframed in ways that relate most directly to the students' level of experience, including cultural background. It is also important to remember not to underestimate the students' level of interest and intelligence. Now that the world of students has been broadened (but not necessarily deepened) by their hours spent surfing the Internet, the interests of some students may be more current than those of their teachers.

What follows may appear to be a restatement of what was covered in the previous chapter. Indeed, this is partly correct. But the review has two important benefits: First, as many of the concepts may be new to teachers, the review will lead to a better understanding. A long-standing friend responded to his first reading of my chapter on Bateson's ideas on the nature of language and double bind thinking by claiming that he was left in a state of confusion. But following a second reading, he acknowledged that he now understood the world in an entirely different and more useful manner. Repetition, when it comes to a radically a new way of thinking that is required in making a transition to a new paradigm, is vitally important. The second reason for the review is that the pedagogical and curricular recommendations are tied directly to the different ecologically and culturally informed concepts. This will enable teachers to avoid making pedagogical and curricular decisions based on vague ideas and the uncertainty of how to help students recognize the connections between their lived cultural patterns, the silences and misconceptions inherited from the past, and ecologically sustainable ways of thinking and cultural patterns.

Teacher Decisions That Highlight the Relational World in Which They and Everything Else Live

Key Understanding: ***Everything in this world exists in relationships within its environment, which is an ecology of interacting relationships and information exchanges that exist within other ecologies. To make the point in a different way that students may be able to understand more easily: There is nothing in nature or in culture that exists in isolation and is totally independent in meeting its basic needs. (It may be necessary to explain what nature and culture mean. It is further suggested that the word*** **culture** ***be used rather***

than* society, *as culture encompasses more of the world that will be examined later.)

How to Share This Understanding with Students:

1. Ask students to identify examples of what they think exists as discrete or autonomous entities. That is, what are examples of things, objects, and, for older students, facts that have their own separate existences?

2. Ask them to describe what exists in relationship with what they may have previously identified as having a separate existence. Also, have students observe how interactions lead to changes in what they identified as having a separate existence. For example, how does water (or the lack of it) affect the behavior of a plant; how do sounds, smells, and even silence affect the behavior of the student; how do changes in temperature lead to changes; and does the appearance of a friend alter how the student thinks and behaves? There are other changes in local contexts that introduce changes in relationships and thus behaviors.

3. Introduce the concept that how we name things and think about them creates the false idea that things, objects, acts, ideas, and so forth have a separate existence from their immediate surroundings. In the case of ideas, ask them to consider whether ideas, facts, or data exist on their own or are dependent upon someone's thought process. There may be an opportunity, given the experience of the students, to ask if language plays a role in representing something as an independent object, thing, or idea—that is, as existing in an unchanging state. Ask why the use of the spoken and written word seldom gives a full account of the changes occurring in local contexts. Are spoken and written descriptions based on someone's interpretation, and what are some of the influences on a person's interpretative framework?

4. Ask students to discuss how their feelings, moods, sense of trust, excitement, and pleasure change as a result of relationships occurring in their environment. Discuss how the changes in their social relationships and in the natural environment affect their sense of identity and self-worth.

5. Ask the students to give close attention to how their patterns of behavior affect the behavior of friends, strangers, animals, and so forth. Older students should be asked to consider how different technologies foster the

illusion that we are separate entities, and thus not accountable for what our behavior communicates to others.

6. This discussion of the relational and interactive nature of life-forming processes should be expanded upon as students move into learning about biology, history, economics, other cultures, how humans are affecting the behavior of natural systems—and how changes in natural systems are affecting humans in terms of how they think, their health, and their general well-being.

Teacher Decisions That Provide an Understanding of the Complex Nature of the Language/Thought Connections

Key Understanding: ***The words we use to name the world and relationships have a history, and as metaphors they carry forward the patterns of thinking and values that existed at an earlier time. Words, in effect, mediate (amplify and reduce) awareness of our relationships with all life-forming and -sustaining processes—misrepresenting some, hiding others due to the lack of an ability to name them and thus to be aware of the patterns, in carrying forward earlier misunderstandings as well as useful insights.***

How to Share This Understanding with Students:

1. Discuss the students' ideas about how they use words, and how their use of words influences their relationships.

2. Introduce the idea of communication as sending ideas and information to others. Explain that this is the conduit view of language; that is, a sender/receiver view of language that hides how the metaphorical nature of words influences awareness of relationships. Also, mention that words, as well as nonverbal patterns of communication, have different meanings in different cultures.

3. Discuss what this view of language hides, such as words having a history that carry forward earlier ways of thinking, and that the speaker's or writer's taken-for-granted way of thinking influences which words will be used to represent what is taken to be reality. Taken-for-granted patterns of thinking and thus the use of vocabulary often reflect differences in life experiences, such as those of a farmer, doctor, pilot, artist, athlete, politician, and person from another culture, and so forth. Ask students to collect the vocabulary of different social groups, including careers, and

ask them to consider how different vocabularies undermine the ability to reach shared understandings and values.

4. Explain the nature of metaphorical thinking where something new is understood as like or similar to something that is already familiar, such as thinking of a forest as like a resource, a new idea as like turning on the light switch, a computer as like the human brain and, now, the human brain as like a computer. Have students identify other examples of analogic thinking, including examples from the different cultures represented in the classroom.

5. Introduce that the analog or analogy that is chosen and that frames the meaning of a word often reflects the taken-for-granted cultural assumptions of the person or group that succeeds in getting later generations to accept the analogs they have chosen. This understanding is important to recognizing that the meanings of words have histories. Discuss the analogies that earlier generations took for granted in their uses of the words *woman, environment, Indian, American,* and *New World.* A key point to emphasize is that the selection of analogies that frame the meaning of words involves relying upon earlier ways of thinking, including deep cultural assumptions, for understanding today's world—which the earlier thinkers did not understand. This is Bateson's point that the "**map**" (metaphorically based language influenced by earlier ways of thinking) does not provide an accurate account of the "**territory**" (that is, today's world). Have students discuss how a road map reflects the thinking and values of the map maker and thus may not represent important features of the territory. This is an analogy for understanding how past ways of thinking continue to influence today's thinking—which is often mistakenly represented as original and cutting-edge thinking.

6. The analogs that frame the meaning of words, and are carried forward when it is overlooked that words have a history, also encode the values of earlier times, such as how the word *woman* encoded the moral values that sanctioned putting her down if she displayed qualities that were not associated with being a proper woman, such as assertiveness, intelligence, independence, creativeness, and so forth. The metaphors that encoded stereotypical thinking also encoded the moral values that sanctioned the mistreatment of different cultural groups, just as thinking of the environment as a natural resource sanctioned the clear-cutting of forests, the removal of the tops of mountains in order to extract the minerals, and so forth. Have students identify other words whose analogs sanction what constitutes moral behavior even though the behaviors are environmen-

tally destructive. *Private property* would be a good example, as both *private* and *property* are based on long-held culturally specific assumptions that ignored environmental issues and that the "individual" is influenced by both the natural and cultural ecologies. Also, discuss examples from different cultures.

7. The meanings of words change over time. This can be seen in how the analogs used to frame the meaning of words previously used to name our enemies have changed now that we are on friendly terms. Also, as we become aware that the human-centered view of the world must be replaced by an understanding of our relational and interdependent existence within natural and cultural ecologies, ask students to consider if metaphors such as *progress, individualism, sustainability,* and *wealth* should take on new meanings more consistent with living more lightly on the land.

8. In the higher grades, students should be engaged in a discussion of whether the use of English nouns represents the relational and thus emergent nature of life processes or a world of static entities fixed in time. Does the use of verbs more accurately represent life-forming processes? Which leads to thinking in abstractions where context is ignored and the abstraction is too often transformed into a culture-free universal truth? Do the nouns used to label a person as having certain qualities take into account that she/he may change in the future?

Teacher Decisions That Present Students with a More Complex Understanding of Print as a Technology

Key Understanding: ***The technology of print has many important uses, but among its primary limitations is that it misrepresents the emergent, relational nature of both natural and cultural ecologies. In short, it reinforces abstract thinking rather than an awareness of ongoing interactions and information exchanges (or semiotic systems) within and between the patterns that connect. It also reinforces the misconception of the individual as an objective observer and thinker.***

How to Share This Understanding with Students:

1. Have students engage in a conversation or other type of activity, with one student being assigned the task of providing a written report of what occurred. This can be the starting point for discussing the important contributions of print, such as providing a written record or someone's personal reflections on the event. The experiences of the students can also be the starting point for a discussion of the limitations of print, such as its in-

ability to accurately represent the dynamic, emergent, interactive nature of the experience.

2. Other experiences, such as watching the crashing of the surf, walking through a forest, or participating in a play or a game can be used to make explicit what is gained and lost when relying upon print. Are printed accounts more accurate than personal perceptions and memory that may be distorted by something threatening in the experience? Can print correct the accuracy of communal memory that is in denial of wrongs done to others? Can print also be used to erase from history environmentally and culturally destructive practices? Have students identify examples of both.

3. Have students read a printed account of a historical or current event. Ask them if the printed account tells anything about the personality and belief system of the person who wrote it. Is this important? Also, ask students to judge whether the written account is an example of how print reinforces a conduit view of communication. That is, does the printed account ignore that most words are metaphors, and thus have a history of reproducing earlier forms of thinking? Examples can be taken from books, including textbooks and educational software, ranging from literature to scientific reports. Printed accounts in newspapers and accounts that are read by television reporters can also be used as examples of how print hides that words have a history, which the reader must remember.

4. Discuss the differences between print and the spoken word. What are the differences that students experience? Following the discussion, ask them if print introduces a form of power that is different from the forms of power exercised in face-to-face communication. Is power being exercised when agreements such as treaties and legal documents are put into print? A further question to explore with older students is: Does the print-based exercise of power take into account the complexities and dynamic nature of local contexts?

5. Ask students what there is about print that reinforces the sense of being an individual thinker and actor. Which form of communication, considering cultural variables, makes it more possible to exercise critical thinking? Are the abstract accounts of an event essential to thinking critically? Is critical thinking in the West based on the taken-for-granted assumption that change always leads to progress?

6. Why do so many people in the dominant culture assume that print provides objective knowledge? Why is face-to-face communication often re-

garded as providing less reliable information than what appears in print? What mode of communication, print or face-to-face communication, provides more information about relationships? Are printed accounts totally free of subjective and cultural influences? Is print inherently ethnocentric—that is, an example of a more evolved form of intelligence and communication? Which is the more cultural colonizing mode of communication: oral or print-based cultural storage and communication? What has been the role of print in establishing ownership of the land in the settlement of America?

Teacher Decisions That Bring Into Question the Myth That Technology Is Simply a Tool, and Whether It Is a Useful or Destructive Tool Depends on the Person or Group Using It

Key Understanding: ***Technologies, whether social or mechanical, are not culturally neutral. While they influence every aspect of modern life, we have little understanding of their cultural and experiential amplification and reduction characteristics—including who benefits and who loses. Development of new technologies is driven by the deepest taken-for-granted assumptions in the dominant culture. The three most important issues related to understanding the cultural non-neutrality of technology are: What is meant by the amplification and reduction characteristics of technology? What are the cultural assumptions that lead to associating technological innovations with progress? Aside from the constructive uses of digital technologies by environmental scientists, how are digital technologies undermining communities and the prospects of a sustainable future?***

How to Share These Understandings with Students:

1. (This issue can be introduced to younger students.) Have students interact with different forms of technology—using a cell phone, writing with a pencil, riding a bicycle or in a car, participating in an activity organized as an assembly line, following a printed recipe, taking a machine-scored test—and ask them to describe in terms of each technology what is amplified in their experience, that is, what enables them to do something that is aided by the nature of the technology. Also ask them to describe what aspects of the experience were reduced or limited due to the nature of the technology. What, for instance, is reduced (lost) in using a cell phone, and what does it amplify (make possible)? What is lost in terms of human experience when riding in a car, and what is amplified (or gained)? How does an assembly line, organizing activities in terms of a mechanical clock, and so forth, amplify certain outcomes while reducing other aspects of experience? Is craft knowledge and skill reduced in

an assembly-line approach to production? Have students identify other technologies and discuss what is amplified and reduced. In the upper grades, this could lead to asking what the use of genetically modified seeds amplify and reduce in terms of the cultural and natural ecologies into which they are introduced. Exploring the amplification and reduction characteristics of social and mechanical technologies will reinforce awareness of emerging relationships and the patterns within both cultural and natural ecologies that connect. Before leaving this topic, ask students to identify what they think are the characteristics shared by mechanical and social technologies.

2. (The following questions are more appropriate for older students.) Ask students why they think most people, including the scientists and engineers who create the technologies, seldom consider what traditional forms of knowledge and skill will be displaced by the introduction of the new technologies. For example, what were the deep, taken-for-granted cultural assumptions held by the people who replaced human skills and control of the pace of work with machines in the early stages of the Industrial Revolution? Have the students list the assumptions that were held then and consider whether these assumptions are still taken for granted by the computer scientists and engineers who are introducing robots into the workplace. Ask more advanced students if these assumptions can be understood collectively as an ideology. What are the connections between a scientific theory such as Darwin's theory of natural selection and how progress is understood? What cultural assumptions are reinforced by the extension of Darwin's theory as a way of explaining cultural changes—and which are best suited to survive over time?

3. What are the characteristics of digital technologies that undermine the intergenerational forms of knowledge and skills that enable people to live less money-dependent and thus less toxin-dependent lives? (Keep in mind that digital technologies are important to gathering data on changes in the world's natural systems, and also keep in mind that print and other abstract systems of representation are the primary means of human/machine communication.) Also, do digital technologies, which rely upon print, reduce awareness of the information pathways within cultural ecologies, including awareness of what natural systems are communicating about the changes they are undergoing?

4. A number of studies are now reporting that nearly half of the jobs in the United States will disappear in a few years as a result of the further computerization of the workplace. Ask the students to discuss what they re-

gard as the most fundamental issues that are not being considered either by the computer scientists whose efforts are making workers redundant or by the employers who can only sell their goods if people have the jobs that enable them to be consumers. Also ask students to consider whether they think democracy is still relevant within the context of the digital revolution.

Teacher Decisions That Introduce Students to the Nature and Community/Ecological Importance of the Cultural Commons

Key Understanding: ***The cultural commons that exist in every community are based on the intergenerational knowledge and skills that enable people to be less dependent upon consumerism and thus live less environmentally disruptive lives. The knowledge and skill are intergenerationally renewed through face-to-face communication and through mentoring relationships.***

How to Share These Understandings with Students:

1. In order for students to recognize a basic distinction that can be introduced in the early grades, ask them to identify the activities, relationships, and things they enjoy that do not cost money. After a discussion of what students have identified, which may vary in terms of ethnic and gender differences, ask them to make a list of activities, relationships, and things they enjoy that must be purchased.

2. Depending upon the experience of the students, introduce students to a list of activities, skills, and relationships that are examples of the intergenerational knowledge and skills that make up the cultural commons. These include food, healing practices, stories, ceremonies, creative arts, craft knowledge, games, knowledge of the behavior of local ecosystems, language, patterns of mutual support, and civil liberties. Also list intergenerational traditions that exclude or exploit others.

3. Have the students interview the parents and grandparents, as well as other members of the community, about some aspect of the cultural commons that they view as now being lost. Do they view technology and the emphasis on consumerism as undermining important traditions? What traditions are being lost, and why were they important? Suggest that students learn a skill or practice that is on the verge of being lost because of the emphasis on deriving knowledge and learning skills online.

4. Suggest that students create a map of the community that shows where the mentors in various aspects of the cultural commons are located—such as master gardeners, weavers, artists, people who practice a craft,

storytellers, and so forth. Part of the mapping process should include interviews that help to identify the mentors' interdependent relationships with others in the community. For students in later grades, exploring the relationship between the mentors' happiness and their physical health might yield important insights about another aspect of the cultural commons that is seldom considered.

5. Initiate a discussion of differences in experience between engaging in a cultural commons experience that is face-to-face and intergenerational and a consumer or work-setting experience where money is the dominant concern. Which leads to discovering a personal talent, skill, interest, and sense of self-worth? And which leads to the feeling of having only a monetary value to others—and to overlooking personal qualities and talents?
6. Initiate a discussion of those aspects of the cultural commons that need to be carried forward and intergenerationally renewed, as well as those aspects that need to be changed or eliminated as sources of injustice.
7. (For older students) Ask students about the characteristics of the cultural commons that make them less vulnerable to cyber disruptions and attacks. Also, ask which aspects of their cultural commons do they see being adopted within the dominant culture—such as food, music, narratives, games, language, and so forth. Also discuss what they think about how the increased reliance upon digital technologies is affecting the oral traditions that carried forward the skills and knowledge of their cultural commons.
8. Given the likelihood that the computerization of the workplace will lead to a higher level of unemployment, ask students to discuss how the traditions of the cultural commons that still exist could be revitalized in ways that would reduce the need for participation in the money economy that will become more limited as the computerization of the workplace moves ahead.

Ask students about the characteristics of the cultural commons that make them less vulnerable to cyber disruptions and attacks.

Teacher Decisions That Introduce Students to the Nature of the Cultural Forces "Enclosing" What Remains of the Local as Well as the Diversity of the World's Cultural Commons

Key Understanding: ***Three of the dominant characteristics of modern society include the many ways innovation and technological change are understood as a progressive force, the culturaly transforming influences of technologies (especially digital technologies), and the emphasis on leading a consumer-dependent existence. In other words, the deep cultural assumptions about progress, technology, and consumerism are powerful forces that are undermining the intergenerational knowledge and skills of the cultural commons that exist in every community and that vary from culture to culture. This process of transforming the gift and largely nonmonetized economy of the cultural commons into a money-dependent and profit-oriented economy is the process of enclosure. With the current drive of global market forces to make workers redundant by adopting computerized machines and outsourcing work to low-wage regions of the world, the challenge is for students to recognize that the cultural commons may represent the only lifestyle available to them that does not reduce them to impoverishment and a state of depression.***

How to Share These Understandings with Students:

1. An important distinction that clarifies what the process of enclosure involves is the difference in the kind of relationships the students (and later as adults) will experience. The cultural commons of interdependent and mutually supportive relationships often involve mentors helping in the development of personal interests and talents. Most experiences dominated by monetized relationships (where one is a worker or a consumer) often require suppressing personal interests, accepting one's place in hierarchical relationships that are often competitive, and performing routine tasks in order to earn a salary. Students in the upper grades should be assigned the task of interviewing workers that range from fast-food employees to office workers in white-collar jobs and even professionals. A question that may lead to important insights would be to ask the workers if they are looking forward to retirement (which might be looking forward to a life engaged in cultural commons activities). Also ask if they

understand the range of activities and mutual support systems that are part of the cultural commons in their community.

2. A major problem is that while everyone participates in the cultural commons of their family, community, and cultural group—and even in many aspects of the cultural commons of the dominant culture—few people are conceptually aware of the differences between cultural commons and consumer-dependent experiences. Thus, in not being aware of the importance of their cultural commons, they are not aware of how it is being enclosed (that is, how the largely nonmonetized relationships and interdependencies are being monetized). A question to pose is whether the dominant cultural assumptions that underlie the industrial (now digital) economy are responsible for why most classroom teachers and professors have ignored introducing students to the ecological and community importance of the cultural commons. As the word *commons* is coming back into use, ask students to visit the web site of the Digital Library of the Commons and to take note of all the studies of the commons and how they refer primarily to how cultures deal with the natural commons. Students should note that there is only a handful of references to the cultural commons, even though it is this aspect of the commons that will become increasingly important as the digital technologies are reducing the need for workers.

3. The earlier discussion of the cultural non-neutrality of technologies, especially digital technologies, should be reviewed and then used as a departure point for discussing how technologies are undermining oral and mentoring traditions that carry forward the knowledge and skills of the cultural commons. Questions to raise: How does print alter awareness of local relationships, behaviors, and contexts? How does it influence the sharing of intergenerational knowledge and the mentoring of skills? Does print, for all its important uses, contribute to the enclosure of the cultural commons—that is, do the abstract representations of reality lead to ignoring the everyday patterns that are sources of information about relationships? What are the differences between information and instructions found on the Internet and learning from face-to-face relationships? Which is less dependent upon participating in a money economy, which will be disappearing for more people as wealth continues to be accumulated by the top tier in society? How do computerized machines or the purchase of ready-made objects contribute to the enclosure of the cultural commons? What are the effects of industrially produced food,

entertainment, games, etc., on the cultural commons? Are they forms of enclosure celebrated in the media as further expressions of progress?

4. The vocabularies inherited from the past are also the basis of ideologies that provide an explanation of current activities and relationships, as well as serve as guides for future political decisions. The question to ask is: How do they influence awareness of the local as well as global cultural commons? Do vocabularies, which might include words (metaphors) such as *progress, individualism, competition, success, tradition,* and so forth, have the same meanings in different cultures? And if not, does an ideology based on the vocabulary of a different culture become a source of cultural colonization? How does cultural colonization impact the viability of the local cultural commons? How does a vocabulary based on an abstract and supposedly rationally based theory, such as the theories of Western philosophers such as John Locke, René Descartes and Adam Smith, affect today's understanding and renewal of the local cultural commons? If the vocabulary provides only a negative way of thinking about the nature of cultural traditions, the practices of the local cultural commons, and the cultural differences in cultural commons practices, will it contribute to the various forms of enclosure of the local cultural commons? To ensure that this discussion is grounded in the experience of students, ask them to identify the key vocabulary that underlies market liberalism and libertarianism, as well as the vocabulary of environmental conservatives. Which will contribute to the intergenerational renewal of the largely nonmonetized relationships and skills of the cultural commons? Again, it is important to continually remind students that language (and thus ideologies) are about relationships, and that past ways of thinking (including the misconceptions and silences) often limit awareness or lead to misinterpreting what is being communicated through the relationship's information pathways. A good example of this is the recent request made by politicians in Texas for scientists to provide a scientific explanation for climate change, but it should consider only the effects of changes on the surface of the sun on global weather patterns—and not the impact of human behavior.

Teacher Decisions That Introduce Students to an Awareness of the Toxic Chemicals That Affect Both Human Health and the Viability of Natural Systems

Key Understanding: ***All cultures have had to learn how to live with the chemicals in their environments. Reliance upon scientific discoveries of the***

chemical basis of life processes has led to many improvements in the quality and extension of human life. At the same time, this same quest to understand and improve upon the chemical properties of what is industrially produced and consumed has led to turning the earth into a sea of toxicants that now affects all forms of life—including humans, who are increasingly burdened with chronic illnesses and physical deformities. The challenge is to introduce students to the basic understanding that the quality of relationships within the cultural and natural ecologies is being impacted by the growing dependence upon the industrially produced toxicants that are now part of the fabric of life. This will be a daunting challenge, but the following represents a starting point for alerting students to a silence that is leading to increasingly devastating consequences.

How to Share These Understandings with Students:

1. The pattern of thinking that can be reinforced even before students are able to understand the properties of chemicals is the same one that has been discussed earlier. Instead of thinking of the properties of things such as individuals, rules, ideas, values, plants, material objects, and so forth, understanding should be focused on the nature of their relationships within the larger world of cultural and environmental ecologies. In the early grades, ask how words affect relationships as well as hide or make explicit what they are aware of. Does relational thinking need to be brought to an awareness of the chemicals in their environment? The starting place is to identify labels and signs on various products, and to think about the relationships that will be changed if the warning signs are ignored. Warning signs and labels are part of a language system that communicates about relationships—but in terms of toxic chemicals, the relationships may lead to illness and even to death.

2. Another basic understanding to get across is that everything that is manufactured is based on the use of chemicals that influence the color, smell, taste, life of a product, its usefulness in carrying out a household task, and so forth. Discuss with students what happens when the chemicals are released into the environment as the products are discarded—including how the chemicals interact with other chemicals in the water, landfill, and in the environment.

3. Interview members in the community who still have knowledge of how pest problems and other issues were resolved before the rise of the chemical industry and the idea that there is a chemical solution to every problem. This could include documenting traditional approaches to healing that relied upon the knowledge of the medicinal characteristics of plants

and other substances, such as the use of honey to heal burns. Also, the interviews could lead to documenting traditional knowledge of caring for animals, as well as knowledge of how to use natural resources, such as ladybugs, to control unwanted invasive species.

4. Introduce students to thinking relationally—that is, ecologically—about how the patterns connect, including, for example, how the industrial uses of chemicals to stimulate the growth of animals, such the use of antibiotics in the raising of chickens, leads to the chemicals eventually being ingested by humans. The patterns that connect, in this example, involve the antibiotics being passed on in the eating of the chicken, and also in the chicken's excrement that is used as fertilizer for the growing of other foods. In this example the antibiotics pass through these different ecologies in trace amounts and end up creating resistance to antibiotics they are used in treating illnesses or injuries. Learning the cultural commons knowledge of different groups' approaches to raising animals that did not require the use of chemicals such as steroid hormones will be useful as more people begin raising their own chickens and relying upon local producers of food. Before inviting the local science teacher to provide further examples of the dangers of industrial approaches to raising food and to healing practices, it is important to review with students the interactive and interdependent nature of cultural and environmental ecologies. The important point is to promote an understanding that there are no self-contained entities, or plants and animals. Everything undergoes change in this interactive and relational world of ecological systems. This applies to the chemicals put onto our bodies, spread across the lawns, thrown into the garbage, and emitted from the exhaust systems of our cars.

Teacher Decisions That Introduce Students to the Emerging Commons That Promote Less Dependence Upon Consumerism: The Makers and Craft Commons

Key Idea: ***The emerging cultural commons of the Makers and the innovative Craft movements offer alternatives to the past traditions of purchasing what has been industrially produced. Both represent a new form of the commons, where ideas and inventions are shared in an open-source Internet-connected environment. What these two emerging cultural commons share with the intergenerationally connected cultural commons that exist in every community is that they are part of a do-it-yourself movement spreading across North America and Europe that focuses on encouraging the development of self-interest and innovative skills that can be shared on the Internet with oth-***

ers, and thus represent a shift way from consumerism. While the Internet enables new online communities to form and share interests, those communities are less dependent upon face-to-face relationships and forms of interdependencies. They also lack the ethnic differences found in face-to-face cultural commons. The Makers are more focused on small-scale technological innovations and projects, and encourage the nonmonetized sharing of ideas. They occasionally hold Makers Faires in different communities across the country. The Craft movement has many of the same characteristics of the Makers, but is more traditional in that the focus is making things out of cloth, leather, and other ready-at-hand materials. It also relies upon sharing techniques and use of materials over the Internet. Both have their own online magazines: **Make** ***magazine (http://makezine.com) and*** **Craft** ***magazine (http://makezine.com/craftzine).***

What Older Students Should Understand about the Differences:

As these two emerging and largely Internet-based cultural commons focus on specialized interests, they lack the more inclusive mutual support systems found in the intergenerationally connected cultural commons. That is, they do not engage their members in the life-supporting and life-enhancing knowledge, skills, and talents found in most cultural commons that range from knowledge of producing and sharing food to traditional healing practices, the wide range of creative arts, traditions of settling disputes, how to live within the limits of the bioregion, and so forth. Both the Makers and Craft movements are still based on promoting the interests and talents of the individual, and thus should be introduced to students in the upper grades. However, students should also be asked to consider whether the do-it-yourself movement provides a critical understanding of the technological and ideological forces that are enclosing the mutual support systems and intergenerational knowledge within the cultural commons that are ecologically sustainable.

Summary: Suggestions for Teachers Who May Feel Overwhelmed

Given that the fragmented nature of the Common Core Curriculum has no overriding conceptual framework, but is instead a list of concepts and skills that will be the subject of testing that is itself embedded in conceptual uncertainties and the inherent limitations of digital technologies, the suggestions here may appear overwhelming to teachers. That they need to take seriously the dominant cultural and ecological challenges of the day is a lot to ask when the teachers' own educations have not prepared them to engage students in thinking about the cultural roots of the ecological crisis. It needs to be reiterated again that what is being suggested here is not that teachers should squeeze entire curriculum units into the different prescribed segments of the common core curriculum. Rather, significant curricular reforms can take place if teachers make the effort to continually introduce ideas, language issues, and cultural patterns during the intervals between teaching the CCS that will be tested, and in the curriculum they initiate.

My recommendations in the preceding chapter, as well as the suggestions here for presenting the basic concepts, language, and examples of lived cultural patterns, need to be used as a resource. To use them most effectively will require that the teacher has a clear understanding of what the priority issues are, which might be posted somewhere in the classroom or even on a class web page. The priorities might include, for example, "relational nature of life processes," "the language/thought/cultural connections," "print as a technology," "cultural transforming nature of technology," "the cultural commons and forces of enclosure," "taken-for-granted cultural assumptions," and "alternatives to a toxic saturated environment." If these priorities are posted where they can be seen, they can serve as a constant reminder of when core understandings can be integrated into extended discussions of the concepts and skills required by the CCS, and in other segments of the curriculum that teachers initiate.

That few of the teachers' professional studies courses will have addressed what has been discussed in this and previous chapters should not be seen as a problem. The concepts and language introduced here should provide the conceptual starting points for engaging students in examining the cultural patterns that are part of their daily experience. There is no paucity of examples of how everything in the student's environment undergoes changes in response to the different forms and levels of communication. Asking students if this is a world of isolated things, ideas, events, and individuals simply challenges them to rethink a long-held myth in the dominant culture. Similarly, students are continually interacting with different technologies, but the question has not likely been asked before about what is amplified or reduced in their experience and in cultural patterns as a result of using different technologies. To cite another example, all students have embodied experiences in a variety of cultural contexts, but few have been asked to discuss the differences between face-to-face communication and print-based communication. And so forth.

The surprising thing about the curriculum reforms being suggested here is that they involve asking simple questions that others have ignored, about autonomy and relationships, the amplification and reduction characteristics of technologies including print, the activities that do not require money, the implications of recognizing that words have a history, and so forth. The other surprise is that so few classroom teachers—and university professors, for that matter—are aware that we are on the cusp of a life-changing ecological crisis, as well as basic changes in our relationships with others as a result of the digital revolution that is blind to what is being lost. What is suggested here is analogous to what happened when the vocabulary and concepts introduced by feminist writers enabled both women and men to become explicitly aware of the taken-for-granted patterns of behavior and thinking that shaped relationships and limited human potential for centuries. This analogy also helps us recognize how resistant people are to recognizing what they take for granted, especially when these patterns are sources of power and wealth. This refusal threatens not only our future survival, but also the ability to recognize the sources of wisdom that should guide our relationships with each other and the environment. What will be addressed in the next chapter, which is the nature of wisdom that is being increasingly marginalized by the twin myths that data and Darwin's theory of evolution should guide our behavior toward others, is totally missing in the CCS reforms that "lay out a vision of what it means to be a literate person in the 21st century"—to quote again the hyperbole that masks the real purposes behind the CCS reforms.

Chapter 6

Another Shortcoming of the Common Core Standards: Knowledge of Wisdom Traditions

J. Progler

This chapter will address what is missing in how a literate person in the 21st century is represented by the Common Core Standards. That is, what can be learned from the wisdom traditions of other cultures? It is a response to the shortcomings of the Common Core Standards (CCS) currently being adopted in the U.S. under pressure from business and political groups that schools prepare students for the workforce. The CCS is driven by analytical, communicational, and quantitative skills in ways that emphasize individualism and utility in a static and mechanistic world. What the CCS lacks is a way of thinking of the world as interdependent cultural and natural ecologies and processes where "relational thinking" replaces the old conceptual patterns that represented the autonomous individual as at the center of the universe. Introducing students to the world's diverse wisdom traditions will enable them to recognize other ways of understanding relationships and identities—and to recognize how their own culture frames the moral issues that arise in their daily relationships with others and the environment.

The CCS seems to leave open the question of curricular content, providing an opportunity to introduce cultural diversity through surveys of traditional wisdom relevant to thematic content in social and global studies. If we move away from the conventional, rational, individually centered pattern of thinking, and toward a system of values based on relational thinking, we can gain insight from the wisdom that has enabled tribal peoples to recognize and thrive within the limitations of their ecosystems by seeing interrelationships between the human, natural, and spiritual worlds. We may also find insights from the wisdom of ethical traditions embedded in the world religions that provide ways for faith-based communities to think about how to relate with others and the world we share.

Emphasizing relational thinking and traditional wisdom seems necessary because market-based values are ascendant and have become our taken-for-granted way of understanding ourselves and the world we inhabit. Aside from the growing realization that individualistic and mechanistic assumptions are now threatening our ecological survival, the current system of values is making us increasingly unable to recognize the forms of wisdom that may help us to think more clearly, or at least differently, about relationships and how we live in the world. There is a wondrous diversity of wisdom available to us, but it has been marginalized by the dual illusion upon which utilitarian schooling rests: that the world is governed by evolutionary biological and social processes, and that all problems and relationships are easily manageable by data. These illusions, increasingly adopted by scientists and humanists alike, are understood here as an absence of wisdom in education. This narrow view promotes learning for the sake of self-centered individualistic and mechanistic goals devoid of ecological awareness of mutually supportive rather than competitive relationships.

Indigenous Traditions of Wisdom

Since the CCS is oriented toward issues related to acquiring wealth through economic development and providing skills to enter the job market, we will begin our journey with voices from tribal peoples that offer ecological ways to conceive of wealth. Many tribal cultures understand wealth in terms of relationships that are immune from the uncertain economic forces of inflation, deflation, and unemployment. In listening to such voices, even while admitting that they may have their own wisdom, one might feel that they are not practical in this day and age because of current concerns about skills for the labor market. From this viewpoint, it may indeed be practical for the CCS to provide students with the skills deemed necessary to survive in the current economic system. But our economic system is based on exploitation of natural as well as human resources, and too often it creates individuals who feel isolated from one another and from the cosmos. From an early age, we are taught to be possessive and individually centered, to "look out for number one." This neglects the impact that competitive relationships have on our sense of well-being and belongingness to society and nature, which in turn has prevented us from appreciating the different ways that other peoples have conceived of wealth.

Bound by a utilitarian monetary view of wealth, Europeans were unable to appreciate the wisdom of the *kula* ring in the Trobriand Islands. Seen from the Western viewpoint, the Trobrianders' perilous sea journeys to engage in transactions with other islanders had no economic logic, since they appeared to trade items devoid of monetary value. But this missed the wisdom of the practice as a system of gift exchanges to build relationships. It took anthropologists and sociologists, such as Bronislaw Malinowski and Marcel Mauss, to see these practices through local eyes. This wisdom is not limited to tribal peoples, as suggested by O. Henry's classic story "The Gift of the Magi." Perhaps we once had this wisdom, but no longer emphasize or live by it. This indicates a clear absence, if not a bias, in what will be inculcated by the CCS. Our modern economic systems and practices have divided us from one another as we moved from lifestyles based on reciprocity to those beholden to systems of distribution; the former is based on interpersonal relationships and the latter is increasingly impersonal and lonely.

Not all activities can be judged by the economic imperative of efficiency. In place of this limited understanding of wealth, we can try to view it through the lens of relational thinking. One could point to the stone dragging among the Weyewa of Indonesia, which from an economic standpoint is highly inefficient but from a relational perspective is very wise. Rather than using machines to cut, move, and set an enormous memorial stone, a Weyewa elder will draw upon his knowledge, call in his favors, and exhaust his material wealth to inspire the entire community to drag the stone by hand in the "primitive" way. An elder engaging in this activity is, as Maybury-Lewis notes, "doing all he can to widen the exchange networks for future generations to ensure their wealth and prosperity" (1992, p. 69). At the end of the stone dragging, an elder such as Lende Mbatu is rich in relationships: As he observed, "I am not a rich man according to most human reckonings but I am rich in ability and I am rich in knowledge, I'm rich in favors and I'm rich in cooperation with others" (1992, p. 72). Economic activities such as these bring people together in reciprocal cycles of exchange, moving the meaning of debt away from its current association with lonely destitution toward being indebted to one another. Our idea of "breaking even," where there is no material economic gain, is seen as pathological because it "breaks" the cycle of reciprocity.

In this relational framework, where giving and lending are not associated with monetary transactions, relationships with animals also change. For the Gabra of Kenya, camels are not only sources of food, clothing, and transportation; they also serve a purpose in a giving and lending economy. Camels are a form of credit that helps to forge bonds among families and tribal clans, and even to create new relationships with strangers. Such a way of living would be disrupted by acting in self-interest, but for the Gabra this is not only destructive of social relations, it will also lead to the selfish person becoming isolated and lonely. What the Gabra possess is not privately owned; this turns giving into a duty and virtue. As one Gabra elder put it, "Even the milk from our own animals does not belong to us. We must give to those who need it, for a poor man shames us all" (1992, p. 85).

A moral economy based on relationships can be contrasted with the amoral market economy, which tends to prey upon and then blame the less fortunate or those in need. Looking at wealth in terms of relationships might provide opportunities for students to reflect on their own understandings of wealth and to question how associating wealth with money and the market economy might create feelings of competitiveness, envy, and failure, all of which are destructive of relationships. It would also be useful for them to reflect on how nonmonetized forms of wealth are returned in ways that build, rather than fracture, relationships.

In tribal societies, identity is more about social relationships than about establishing oneself as an autonomous individual. An example of this can be found in naming practices. In the Xerente tribe of Brazil, children are not named for several years after birth; they are referred to publicly during that time with general relational terms such as *daughter, son, niece, nephew*. In his fieldwork among the Xerente, Maybury-Lewis (1992, p. 121) concluded that naming was "too important to be casually bestowed on babies by parental whim," because "a name is not a device for singling out an individual; it is a way of making an individual into a social being, of linking him with society, just as society is linked to the cosmos." For the Xerente, as with many tribal peoples, naming provides a social personality. In this way, tribal worlds are more clearly defined and therefore secure, because identity and socialization come through interaction and rely on community support and participation. Native American philosopher Vine Deloria suggests that, from a tribal point of view, the idea of a lone individual is "ridiculous," because "the very complexity of tribal life and the interdependence of people on one another makes this conception improbable at best, a terrifying loss of identity at worst" (1992, p. 123). Industrialized societies provide some guidance for identity, but this is often prioritized to fit political and economic needs, as evidenced by the CCS emphasis on job skills. Guidance for developing a sense of self is left for individuals to cobble together from the destabilizing forces of the media and the market. Conceiving of identity as a process dependent upon relationships may help students to evaluate how institutions and peer pressures influence their sense of self, and to seek in their community's traditions the wisdom to avoid being shaped by haphazard values that few members of the peer group have really thought about.

In addition to socializing students to the values of the marketplace, modern schooling also inculcates the values of the national political system. We often defend our sense of individuality by pointing to the economic and political systems that value freedom and democracy. However, these beliefs have come to depend upon the state, and the state itself has become heavy-handed, so it may be hasty to assume that we have the only viable political system—and that this system is the only way to insure our individuality. Values such as individual freedom and democratic decision making are also found in tribal societies, in which efforts are made to avoid the concentration of power that often accompanies modern political institutions. An especially important example, and one which historians have suggested may have influenced the framers of the U.S. constitution, is from the Haudenosaunee, or Iroquois, nations who reside in what is now New York and Ontario. They negotiated tribal alliances that provided bonds of unity and valued diversity. Tribal chiefs were and are recognized for the qualities of their character, and selected for their eloquent oratory and skill in mediation. This eloquence and mediation impressed even the likes of Benjamin Franklin, who referred to the Haudenosaunee model when encouraging his compatriots to forge an alliance among the British colonies in North America. However, tribal alliances did not vest their authority in a state, which is where the emerging American democracy diverged from tribal wisdom. As oratory is the basis for tribal mediation, after contact with Europeans, many indigenous peoples of the Americas came to see literacy as a vehicle for lies and deception, owing to the countless broken treaties that were initiated, and often enforced, by the colonial powers through manipulation of the written word.

Learning how tribal and other non-Western societies limit the centralization of power can prompt students to seek examples within their own ethnic groups and local communities, especially those that carry forward traditions and that identify the qualities of people who are recognized as sources of wisdom. Teachers might suggest ways to talk about the differences in character that separate community-based leaders from modern politicians who are voted into office, and what aspects of personality and appearance seem suited to our media-driven approach to selecting leaders, and whether these selection processes have anything to do with wisdom.

Tribal and industrialized societies also differ in their views of aesthetics, and these differences have implications for intercommunal relations. Aesthetics can be understood in relational terms as affecting the well-being of others, which can contribute to a broader sense of well-being in the world. Social and natural environments lacking in beauty degrade our sense of well-being. While those of us in the industrialized world tend to view art as an expression of the autonomous individual or as something to be displayed in museums, for tribal peoples aesthetics retain an association with a sense of beauty that is extended beyond creating material objects to living beautifully. A profound example of this view of life as a work of art is found among the Wodaabe of Niger, whose culture revolves around living a life of beauty and grace, in social relations as well as celebratory occasions. Their sense of aesthetics deepens their relationships. For the Wodaabe, being beautiful includes the physical dimensions of grooming and clothing, but it hinges upon the social virtue of charm and the ability to interact with others, which in turn accompanies other social virtues such as patience, loyalty, and reserve. There are no artists in Wodaabe society; everyone lives life artfully.

Being primarily oral-based, tribal societies put a high value on memory, which is beneficial in ways that we may not recognize because of our fixation on literacy. In our world, we see orality as the absence of written texts, and that people without writing are "illiterate." Ironically, written texts are dead letters, frozen in time and space, while oral speech is dynamic and alive. For tribal peoples, speech bolsters community because of its inherent dependence on face-to-face communication.

We tend to rely less on memory, and our educational systems have often denigrated it in favor of literacy—and now data-storage technologies. The emphasis on utilitarian skills in modern schooling reconfigures memory as "rote memorization," which in turn is seen by constructivists as a hindrance to learning.

Reliance upon print fosters the illusion of being an autonomous individual who engages in critical thinking—which is largely influenced by the form of intelligence encoded in the language that she/he takes for granted. For tribal peoples, memory is a community activity which is dependent upon relationships. We might also feel that human memory is limited and therefore that our books, libraries, and digital archives are somehow more reliable ways of knowing. But tribal peoples disperse knowledge throughout the community by way of memory, and they tend to remember what is important or necessary and disregard the rest. In our world, dominated as it is by texts and computers, we have lost the ability to determine what is important and instead are drowning in an ever-expanding sea of data. Understanding memory as a communal activity can encourage students to explore the relevance of orality and memory in their own communities, through which they might collectively identify wisdom among ethnic groups in American society. Teachers can ask students to consider how memories, such as those associated with appropriating the land of indigenous peoples, or keeping other peoples in bondage, or maintaining status based on wealth and class, can be rectified, and which sources of community-based wisdom might guide such efforts.

In addition to encouraging community relations through collective memory, orality amplifies the spiritual awareness of tribal peoples. Without texts and computer screens to colonize experience, the senses can be opened to other channels of awareness that can heighten relationships with nature and with the world of the spirit. Whereas in industrialized societies it is often taken for granted that there is a separation between spirituality and everyday life, in tribal societies the division is less evident. Tribal peoples see the world as sacred, which differs from our modern cultural assumptions that proceed from a desacralized world. These viewpoints impact how we understand one another and the world we share. For tribal peoples, the sacred is reinforced by rituals that nurture ecological relationships, connecting people to one another and to nature, and spirituality pervades life in ways that we cannot recognize because of our tendency to categorize experience into sacred and profane, religious and secular. Many tribal societies do not even have a word for religion, although their lifeways can be profoundly religious. This absence of the spiritual dimension of life, symptomatic of the CCS and U.S. schooling in general, undermines our imaginations and dreams, and is implicated in the disenchantment of the world. Listening to other peoples for whom these capacities remain intact might help to re-enchant our world by seeing connections to realms beyond ourselves and our machines.

From his work among the Xerente, Maybury-Lewis found that developing spiritual awareness involved adherence to the traditions passed down by the society, with meticulous attention to the details of ritual that might seem unimportant to our eyes, but when examined with care and open-mindedness bring clarity to crucial features of the tribal outlook. Margaret Mead noticed this long ago, when she asked children during her fieldwork in Papua New Guinea to draw pictures and then noticed that they did not depict the spirits and other supernatural entities that were an important part of adult life. She concluded that spiritual awareness is not innate, but is taught and reinforced by society. We might see this as a triumph of the secular view that religion is merely a form of social indoctrination, but we can also see it in terms of what might be absent from our lives, and by extension, what is absent from our educational imperatives.

It is not necessary to live with tribal peoples to experience and understand their outlook on life. A common practice among many tribal peoples is the sharing of dreams each morning. These are not dreams as embodiments of hopes and aspirations, or as a world of symbols indicating repressed desires from waking life, as we have come to believe. Rather, the tribal sharing of dreams is about "where we go while asleep." These dream journeys, which might appear fanciful for us, take on a deep significance for tribal peoples, and the daily communal sharing of dreams forges relationships with one another and with the spirit world (Maybury-Lewis, 1992, p. 206). This suggests that for tribal peoples, dreams are "opportunities for pilgrimages of the soul," and these pilgrimages provide inspiration for cultural activity in the waking world.

The experiential dimension of spirituality, which is pervasive and favored among tribal peoples, is not absent from the historical religions. However, this dimension is relegated to the fringes, with the division being most noticeable where religion has collapsed into doctrine. But it is in precisely these areas that we find some of the most important evidence of community and spirituality, even in the modern industrialized societies. For example, Sufism, the mystical branch of Islam, rose to prominence in the West from the 1960s, and is usually seen as a safe form of Islam for weekend spiritual wayfarers because the focus is on the experience of the Divine rather than on correct doctrine. For the mystical branches of the historical religions, the goal is to experience the mysteries rather than merely believe in them, and these experiences often take on a communal dimension in the practice of rituals and collective prayers.

There are precedents in the modern societies that provide a way for us to appreciate, or even comprehend, what tribal peoples are taking for granted. What is shared by the mystics of civilization and tribal societies are the communal bonds among humans engaged in the spiritual quest, and between humans and the spirit world. In terms of education, perhaps we owe our students, and ourselves, an acknowledgement that there are other worlds beyond the mechanistic and utilitarian, and that the difference that makes a difference in terms of living and thinking ecologically might reside in our inability to experience the world as mysterious and sacred. This would seriously reduce the tendency to think of the world only in terms of its economic value.

The priests of the historical religions and the scientists of the industrialized societies are systematically trained to recognize what ordinary people are unable to comprehend. For tribal societies, the shaman takes on a similar role, mediating between the seen and unseen worlds, and often performing healing, both individual and communal. During his fieldwork among the Azande of Central Africa, Evans-Pritchard noticed that a shaman had interpreted from a spiritual as well as a counseling point of view a mishap in the village that involved the collapse of a granary and the death of a man sleeping beneath it.

Whereas those schooled in the West might ask *how* this happened, the shaman asked *why* it happened. These two deceptively simple questions reveal different outlooks between tribal and modern societies that make a difference in what is regarded as important information. While both are concerned with understanding what had occurred and preventing it from happening again, asking *how* it happened may lead to seeing the event as a *coincidence,* which is the term scientists use to dismiss what they cannot explain. This might usefully lead to recommendations to build a better granary, perhaps more impervious to wind or termites. While the Azande do not disregard such practical knowledge, asking *why* led the shaman to focus on disharmony in the community: Someone must have placed a spell or evoked an "evil eye" on the person that died. This focus on relationships turned a mishap into an opportunity to re-evaluate and heal community relations. Reflecting on such stories can help students explore the questions asked by scientists and those addressed in different cultural and spiritual traditions, and to further consider the implications of separating questions of facts and data from questions of morality, especially in terms of the morality that impacts how we understand and relate to one another.

Sometimes the spiritual and scientific worlds coexist. Physical pilgrimages to sacred places are an important part of tribal peoples' spiritualities, but at the same time, they are also community events. Among the Huichol of Mexico, pilgrimages become a way to align the community with the cosmos, helping to bring relationships into balance, alleviate social strife, heal illness, and rebalance cosmic forces. The *Millennium* television series relates the story of an epidemic among the Huichol that was causing the deaths of many children. A modern doctor, at a loss to cure or explain the illness, resorted to sending samples for laboratory analysis. Adopting a different approach, a shaman noticed that the illness coincided with a fog that enveloped the mountainous village. He realized that their world had gone out of balance due to the Huichol focusing too much attention on the mountain spirits and not enough on the sea spirits, causing the sea spirits to send an illness in the fog. This could only be made right by a pilgrimage to the seaside to make offerings. By the time they had returned to the village, the lab results were back, and they indicated that the children had died from a mutated form of measles, which could then be vaccinated against but not cured. Which led to the healing? It depends on our outlook, but what is remarkable about the Huichol is that they took seriously the belief that worldly events are connected to spiritual realities, without rejecting the medicine of the modern doctor. This story is also as an example of intercultural dialogue. The doctor in this case was unusual; although he was trained at the university to denigrate such practices, he still cooperated with the shaman, who in turn cooperated with the doctor. Both had the same goal: to heal. Their cooperation came about through a sense of mutual respect.

The spiritual and material worlds are connected in other ways that blur the boundaries between social and spiritual relationships. In a remarkable example of "ecology of mind," the Australian Aborigines have a worldview based on the Dreamtime, the spiritual world from which all has emanated and to which all is connected. It is the world of the ancestors, but it is also a parallel world to ours. Frans Hoogland, a Dutchman who lives among the Aborigines, apprenticed himself to a community elder and learned a new worldview that sees connections between humans and the cosmos. He noticed that these connections are enacted through rituals:

> We start with nothing—a total emptiness—a void. Then we have singing and dancing. We start by forming—the singing creates the sound and the vibration forms a shape, and the dancing helps solidify it. The dancing is making the form stand out as a tree, a bird, as land. The process, the Dreaming itself, becomes reality, something we can work with and see. (Maybury-Lewis, 1992, p. 198)

The physical world in which we live is a relational world that assumes an interconnectedness with the Dreamtime, but for the Aborigines this relationship is dynamic. Reality as they experience it is dependent in some way on their actions, and so to keep the world stable, it is necessary to perform rituals—in this case, dancing along "songlines" that form a cosmic map of the terrain. What appears to be a community activity, reinforcing bonds among members with one another, also creates and maintains connections with the unseen world. As Hoogland puts it, "To pay respect to the country, you will, as you walk along, say hello to the places through the song and dancing" (Maybury-Lewis, 1992, p. 248). For the Aborigines this is a "cultivation process" akin to, but not the same as, cultivating the land for food production; rather, it is a spiritual cultivation that creates a bond with the land. Maybury-Lewis suggests that these communal songs and dances are the cosmic rhythms and melodies that give the everyday world its form, but admits that this outlook appears strange to us because we are accustomed to thinking in Newtonian terms. In a world informed by science, he continues, objects are seen to be "separated from one another and only become connected when force is applied to them. The Aboriginal system rejects our separation of the visible world into discrete objects, just as it denies that matter is the primary level of reality" (1992, p. 201).

Understanding how tribal peoples see life as sacred may help students to reconsider how our own deep cultural assumptions categorize the sacred and secular, where these categories come from, and what might be the social and ecological implications of maintaining them. Students can turn to their own communities and those of their peers in observing how rituals and ceremonies might serve the purpose of creating and maintaining relationships and a healthy attitude toward the environment, and which rituals foster well-being within American society, and, conversely, which rituals may contribute to moral relativism, continued social conflict, and environmental destruction.

Faith Traditions of Wisdom

Another area neglected by the utilitarian and mechanistic worldview of the CCS involves the wisdom found in traditions of faith. This overlaps with the above discussion of spirituality, but also relates to our familiarity with the historical religions. Exploring the wisdom of faith may help students to find shared ways of understanding traditional ethical precepts and to reflect on their relevance in today's world. Focusing on ethical dimensions of the historical religions can avoid the distracting debates about the existence of God or the date of creation. Within the Judeo-Christian tradition, we can instead consider the four ethical precepts of the Ten Commandments, which forbid murder, theft, dishonesty, and adultery, and reflect on how they proscribe the behaviors that lead to animosity and societal fragmentation.

It is a modern curiosity indeed that the ethical teachings of this tradition have been overshadowed by theological and historical disputes that do not really help us to get along with one another. Accepting that the religions seem to have recognized some of the destructive behaviors that threaten social relations is not a matter of preaching morality. Similarly, while the Jewish emphasis on social justice has influenced both Christianity and Islam, it is often forgotten that the ethical precepts informed by calls to justice from the monotheistic prophets have provided the basis for many of the legal traditions that continue to inform our secular social systems today, even if their roots in religion are obscured. It has become commonplace when talking about Martin Luther King Jr., for example, to refer to his prophetic voice as a powerful call for justice in the U.S. civil rights movement.

Some of the wisdom traditions, such as those of South Asia, take an approach to life and relationships that begins with the inner dimension and works outward.

In the Hindu tradition, one of the aims of life is to move beyond self-centeredness and embark upon "the path of renunciation," which includes a focus on selfless community service that can be developed through practice of *karma* yoga. Quite beyond the impoverished yet widely commodified notion of yoga in the West as merely a self-centered bodily exercise, yoga in the Hindu tradition is intended to integrate ourselves with the cosmos. As one of the four varieties of yoga, karma yoga is a spiritual technology that proceeds from the belief that every outward act has a reaction on the actor and then attempts to focus the intention of the act away from the selfish benefit of the actor.

In Buddhism, what might be seen as ethical teachings proceed from the Five Precepts, some of which are resonant with the Judeo-Christian Ten Commandments, and which similarly include admonitions against killing, stealing, lying, and adultery. However, rather than culminating in a codified legal tradition, these Buddhist precepts are understood as obstacles on the path to enlightenment. Here the difference that makes a difference is that while they may resemble our laws, for Buddhists these precepts circumscribe acts not because they are *illegal* but because they are *incompatible* with the spiritual quest. This points to how faith has ethical implications for how we treat ourselves, others, and the world.

Perhaps the wisdom tradition that provides the clearest guidance for how people ought to relate to one another is Confucianism. We can put aside the distracting debates on whether Confucianism is a religion or a philosophical or ethical system because we are interested here in its influence as a wisdom tradition, and as such, what it might have to say about seeing and living in a relational world. The teachings of Confucius have become so ingrained over the 2,500 years since he is said to have uttered them that they are now seen as natural and timeless features of East Asian societies. Confucianism seems to have gone beyond explicit faith into implicit culture. It has become a taken-for-granted mode of being. Although Confucius sought to attain a political office to solve the social problems of his day, ultimately he retreated into teaching. What he taught is perhaps one of the most fully formed and widely dispersed ethical views among the wisdom traditions.

While we may see them as "Asian values," most people living in the societies impacted by Confucianism are not concerned with the origin of these traditions. Confucian values have for centuries formed the normative basis for social relations. Huston Smith suggests that Confucian social relations center on empathy, which in public life "prompts untiring diligence" and in private life promotes courtesy and selflessness and "the capacity to measure the feelings of others by one's own" (1994, p. 110). This may help explain why East Asian social relations appear bound up with the group, a type of collectivism that contrasts with Western individualism.

Collectivism and individualism are not necessarily mutually exclusive, even though language conventions make them appear so. Remembering this can help us to reflect on what we are teaching in schools. As Bateson suggested, the conceptual "map is not the territory"—that is, it may be an outdated way of thinking. Whether we are talking about the CCS or the wisdom traditions, it is important to remember that these conceptual maps are often incomplete, transient representations of reality. If we are aware of their limitations, we can perhaps better understand the world we all share. If the CCS is to prepare American children to live in the world, or just to find a job, then as that world becomes more multicultural and complex, it seems prudent to take into consideration values other than those taken for granted in the dominant culture. By promoting individualism, the CCS reforms overlook the possibility that individual behaviors and values can have lasting social implications. In the end, being an individual is still a social act, as much so as collectivism.

Teachers should carefully highlight such differences by identifying the moral and ethical insights of different religions, and even provide some additional background explanations of their spiritual roots while avoiding indoctrinating students. This can be achieved by asking which cultural assumptions make it difficult to recognize the implicit moral values inherent in the way we talk about the world, and how our cultural assumptions, embedded in language, might perpetuate an ecology of destructive relationships. Teachers should also remind students that words are metaphors, that words and ideas have histories, and that present-day meanings continue to be guided by analogs settled upon in the distant past. This could involve unpacking the assumptions embedded in the terms that we take for granted, such as *woman, environment,* or *resources.* By looking at how outmoded analogs inform present-day metaphors, teachers can help students grasp what was morally and ethically permissible at the time these analogs were first adopted, and how they continue to influence our world today.

Returning to China, a crucial component of what scholars of religion call its "great religion"—the intersection of Confucianism, Taoism, and Buddhism—is Taoism, which has perhaps the most sophisticated view of the relationship between human beings and the environment. The Taoist view proceeds from a threefold concept of the Tao as the transcendent way of ultimate reality, extant but largely unknowable through human intellect, to the immanent view of the Tao as the way of the universe through which all life is derived and regulated, to the way of human life that involves living in light of these transcendent and immanent ways, and which hinges on the cultivation of creative quietude. Rather than fighting for dominion over nature, creative quietude might be seen as "going with the flow" of ecological relationships. This value informs everything from architecture, which works with rather than against nature, to social relations, which ought to avoid friction and conflicts, to medicine, which attempts to make use of the body's own natural characteristics for healing. Owing to the view that nature and human beings are interconnected by the Tao, natural medicine has evolved in China and Taiwan that relies upon herbal treatments to realign the body and mind with the ultimate way of the universe. The flow of the Tao in the universe is paralleled with the flow of energy in the human body, which can only be brought back into alignment with the Tao through diet, exercise, and less stressful social relations.

In Taoism, social well-being comes about through the psychic dimension of nonintervention, in which the human ego and consciousness give way to a power beyond themselves while living in ways that avoid stress and excess. Taoist teachers employ water metaphors to illustrate these points. Water is supple and yielding, assuming the shape of whatever container into which it is poured, but it is at the same time powerful in subduing that which is solid and unmoving. The notion of "muddying the waters," often evoked in problem-solving activities, points to the Taoist observation that sometimes the best thing to do is nothing, to remain calm instead of instigating conflict.

Taoism shares with Confucianism an emphasis on humility, but this humility is not limited to the hierarchical social relations of Confucian values; it is a humility in the face of nature. They share an emphasis on living in harmony, but they also diverge in that whereas Confucianism sees acting upon the world as a neutral and even necessary feature of relationships, which therefore gives way to the civilizational impulse to subdue nature for human benefit, Taoism ultimately prefers to leave nature as it is and to instead find ways of aligning ourselves with it. Taoism, as suggested by the well-known yin/yang symbol, represents the interrelationship of values, that there are no absolute opposites, and that opposites are not in competition; they are complementary. There is no strict line between good and evil, either, since in every good act there may be a little bit of evil, and vice versa, which should not be confused with a moral relativism that ultimately becomes amoral or indifferent. There is no immutable line between light and dark, or between the genders, or between other seemingly obvious opposites. Within one lies an element of the other, and both are interrelated to form a larger whole, a harmony.

Given that the world of relationships is dynamic, rather than consisting of fixed entities and permanent connections, we might consider how the wisdom traditions can help awaken the feeling of being participants in an interdependent world, and how this might lead to reflections on life beyond the immediate experiences of the autonomous individual. In turn, this may help us to rethink our proclivities for controlling others, accumulating money and acquiring status, or believing in the myth of progress, and how these beliefs undermine our ability to experience life as a larger ecology of intertwining nurturing relationships. Teachers could perhaps encourage students to reflect on how selfishness, the desire to dominate peers, or being rude, indifferent, or disrespectful to others can lead to an ecology of bad experiences that spiral downward in ways that will affect and inform our future relationships with each other and nature.

In addition to highlighting the ethical dimension of faith, scholars of religion have identified three virtues that characterize a life informed by wisdom: charity, humility, and truthfulness. As with ethics, these virtues have taken on meanings over time that may have obscured or overridden their value in strengthening relationships and the well-being of others, and turned them into means for achieving self-interest. How we treat others is bound up with the important activity of dialogue, and seeing the virtues as examples of relational thinking can help us to move from monologue to dialogue, in terms of how we relate both to the world and to one another. Perhaps the most familiar of the virtues embodied by the wisdom traditions is that of charity. Most of the world religions recommend some form of charity, and communities of faith have promoted both public and private acts of charitable giving. The Islamic tradition even makes charity an article of faith—one of the Five Pillars of Islam, the proceeds of which are redistributed within the community. But it would be mistaken to view this as a poor tax. The transaction, though worldly, is ultimately between believers and God, an act of faith and not a function of economic policy. Charity plays the dual role of helping communities of faith work toward the Divine while encouraging them to redistribute their wealth. For Buddhists, this broader sense of charity extends to ways of speaking, in that lying, gossip, slander, and other forms of verbal abuse block the path to enlightenment—another illustration of how the spiritual is interrelated with the social. In the Confucian world, charity is intertwined with humility and the social conventions of respecting others and putting them at ease.

Teachers could encourage students to examine how other traditions view and practice charity. The contemporary view of charity as an economic transaction from the rich to the poor appears impoverished when seen against the wisdom traditions. Although both forms of charity are evident in social relations today, the motivations differ; one proceeds from self-interest and the other is based on selflessness. Beyond the modern, divisive connotation of "donation," charity can be broadly conceived as a way of viewing others as equally valid to ourselves, as an expression of empathy. This might encourage teachers to consider how empathy is itself an expression of wisdom, which may in turn lead to discussions on whether it is apparent or important in student relationships today, or more broadly, in human-nature relationships. Teachers might also engage students in a discussion of "possessive individualism" and how this modern tendency can undermine empathy.

The virtue of humility is well illustrated by the Confucian tradition. Rather than leaving the decisions about how we act toward others to individual whim, or to state power and civil ordinances, we might focus on how humility informs relationships that reflect on the self as much as upon others. Humility involves elements of what sociologists call "filial piety"; there are proper ways to behave according to relationships, including with the ancestors. Though not religious in terms of being a tenet or article of faith, piety toward one's elders is evident in the well-known Asian value of respecting parents and the aged. In the Confucian world, which extends beyond China to other East Asian societies, humility is normative and ritualized to the point of being taken for granted, much the way individualism is for us. The self in this world is not an isolated entity; it is a node in a system of relationships. While in the West we have come to value the autonomous individual, most may still recognize that even superficially, Asian societies tend to be more group oriented. It is not so important that this has proceeded from the teachings of Confucius. What matters is that these values have become normalized as part of the culture, and it is education, which was at the core of the Confucian project, that has normalized collective values. Again, it's not necessary to prescribe these values. But it is important to recognize their wisdom, which is based on a radically different sense of the self than the one to which we are accustomed. In order for social relations to improve, individuals can try to accept that they are part of larger relationships that are perhaps best envisioned as seeing the self in a series of concentric circles, spreading outward to the family, and to the community, the society, the civilization, and ultimately to the universe. This is a form of self-examining that might help us move beyond a self-centeredness that pervades our relationships today.

Turning to the virtue of truthfulness, we can point to Hinduism and its nuanced view of reality that proceeds from the belief that although the world may appear to be as we perceive it, this appearance is not the true nature of the world. In Hinduism, our ability to see is obscured by "walls of illusion" that are built upon a narrow sense of self, and which prevent us from seeing the world as it is. Thus, in the Hindu tradition there is an emphasis on transcending the self to reach an integrated state of awareness where all is seen as one, giving us perhaps the most relational of all the teachings of the wisdom traditions. In another view of the virtue of truthfulness, Zen Buddhism focuses on the limitations of language to comprehend reality—that the "the map is not the territory," to return to Bateson's phrase. Zen encourages negating the rational truth-seeking mechanisms of the mind to clear the way for a direct experience of reality, leaving aside the map to commune with the territory.

When Huston Smith lived for some time as a monk at a Zen monastery in Japan, he learned this firsthand. The community of monks was brought together under a Zen master who guided each individual according to what he needed to experience this ultimate reality. For Smith, this meant that he had to cure himself of what the master called the "philosophy disease." We don't need to undertake the rigors of yoga or apprentice ourselves to a Zen master to appreciate that these examples point to very different ways of understanding reality and truth. The narrow prioritizing in the CCS reforms of rational truth-seeking mechanisms aided by computers stands in stark contrast to these views of the world, pointing to another way that the standards marginalize the wisdom of tradition.

Ultimately, the wisdom traditions provide us with a vision of the world that scholars of religion suggest consists of unity, optimism, and mystery. For our purposes here, it will be instructive to focus on the vision of unity. While the modern scientific method has led to some astonishing observations, this is largely done by fragmenting and breaking down reality, and by isolating and examining the smallest parts, even for answers to the biggest questions. At times, a bigger picture can be deduced from the fragments, but this is not a requirement for practicing science, and in any case, this is often dismissed as practicing philosophy. The wisdom traditions, on the other hand, begin with the view that reality is more integrated than we may assume. Not constrained by the tenets of the scientific method and the limitations of an empiricism that restricts itself to what is quantifiable, the wisdom traditions work in the other direction by beginning with an integrated view and then looking at the details. This vision of unity as found in the wisdom traditions indicates a form of relational thinking, as Huston Smith eloquently puts it:

> Mortal life gives no view of the whole; we see things in dribs and drabs, as self-interest skews perspective grotesquely. It is as if life were a great tapestry which we face from its wrong side. This gives it the appearance of a maze of knots and threads that look chaotic. From a purely human standpoint, the wisdom traditions are the species' most prolonged and serious attempts to infer from the hind side of life's tapestry its frontal design. As the beauty and harmony of the design derives from the way its parts interweave, the design confers on those parts a significance they are denied in isolation. We could almost say that seeing ourselves as belonging to the whole is what religion—*religio*, rebinding—is. It is mankind's fundamental thrust at unification. (1994, p. 248)

This is not to promote "intelligent design," which subordinates the observation of wholeness to the divisive question that lies at the center of a dispute between religion and science. Bracket out divisive questions and we have a statement of the inherent holism, the interconnectedness of life and spirit, the material world and the immaterial, that transcends the perceived boundaries laid down by the scientific method and the ensuing arguments that proceed from its applicability to questions it was not intended to answer. And limiting the element of design to proving the existence of God is ignoring the other religions, such as those of South and East Asia, in which God does not play a central role.

It is as if our comprehension of reality, as evidenced in the way we understand ourselves, one another, and nature, is mired in the assumptions of the West Asian religions and their antagonists, with both parties to these arguments cornering themselves as they emphasize contradictory points to the exclusion of others. Meanwhile, the dominant arguments themselves have excluded traditional wisdom by focusing only on a small, self-serving dimension of what the West Asian traditions are saying in order to discredit or undermine their other points, while omitting religions that are not configured in the same way as those of West Asia. In fact, the obsession with proving or disproving the existence of God has obscured other aspects of the wisdom traditions and their unitary vision of the world, and has also destroyed the most important skill that we might learn, which is to listen.

But if we are to truly listen to the varied voices of others, we have to also identify and face the obstacles to understanding and empathy. This might involve considering how some of the values and assumptions of modern industrialized society might actually impede learning from the wisdom that can be found in various tribal, ethnic, and religious traditions. For example, by extending the theory of evolution beyond it being a way to explain the fundamentals of biological development to using it to evaluate social and cultural developments, we might actually be displacing important moral insights of different religions and wisdom traditions. Suspending this reliance on social Darwinism might then help us not only to listen to the voices of others, but also to consider and evaluate the guiding moral norms that govern our own social and environmental relationships, and to reconsider what it means to be "better adapted," or what is meant by "survival of the fittest," and how the meanings we ascribe to these terms carry hidden moral guidelines.

In the end, education can be a pilgrimage, a journey of becoming, toward seeing the world anew and experiencing life with a sense of awe and even reverence. It is perhaps unfortunate, and maybe even tragic, that education has been reduced instead to the narrow and utilitarian aims as embedded in the CCS, the implied intent of which is to provide, under the rubric of economic development and progress, cogs for the machine of our transient modern industrialized economy— even if that machine is coming more and more to resemble a juggernaut. It's not that the wisdom traditions can get us off this juggernaut; in some cases, we are happily riding along. But how sad it is that we sentence our children to 12 or more years of this narrow training, especially in light of the wondrous things we have seen in our very cursory journey through the wisdom traditions, which I hope might at least help us to see the world as being more interconnected than we think it is.

At the same time, there is no pretension of having said it all. We have barely peered beyond the surface layers. So rather than concluding something definitively, it is my hope that this journey will begin anew with the call to dialogue as noted above, and that we may take to heart the crucial role of listening needed to sustain dialogue. This sense of listening in dialogue, if I may be so bold as to suggest, needs to pervade what we do as educators, and it does not need to be limited to dialogues among wisdom traditions, which are in any case a good start. Ultimately, we ought to consider engaging in the greater dialogue of what we want to achieve with all this time and energy spent on educating our children and what role the wonders of life may play in it, and what are the implications of exiling that world of wonder from our educational experiences. Standards come and go, institutions wither and fold, but the cultural values that we teach and nurture today will be with us long after the standards and institutions have gone, so we ought to take a closer look at what values we are teaching, what has become our standard of excellence, and how the diversity we've seen can inform where we might go from here.

J. Progler is Professor of Culture, Society and Media at Ritsumeikan Asia Pacific University in Japan and co-creator of the Multiversity project in Malaysia and India.

Chapter 7

Unanticipated Consequences of Making Computer Science Part of the School Curriculum

As explained earlier, double bind thinking, behaviors, and policies often lead to constructive short-term outcomes, and at the same time, to destructive long-term consequences. For example, expanding the economy, which has immediate benefits in terms of increased employment, also leads to more toxins being put into the environment and to the further exploitation of natural resources that are not infinite. Another example still fresh in the experience of people living on the east coast of Canada and the New England region of the United States is how the increased use of technology in the fishing industry led to immediate economic gains in terms of larger catches, but over the long term decimated the codfish fisheries to the point where their numbers went into a radical decline, and still have not recovered. As many students are discovering, going into massive debt in order to obtain a degree from an elite university, which in the past ensured a high-paying and high-status job, has only a short-term benefit of obtaining a degree, until it comes time to find a job when increasing numbers of jobs have been taken over by computers.

The same double bind thinking is also present in the current effort to introduce the coding skills of computer science into the nation's public schools. At first glance, the number of school districts adopting computer science as part of the curriculum, especially the Hour of Code program offered by Code.org, is impressive—with schools in New York and Chicago working directly with Code.org. According to Hadi and Ali Partovi, the two founders of Code.org, 17 million people in the United States, half of them women, have already participated in the Saturday program. Backed by Microsoft, Google, Amazon, the College Board, and the Computer Science Teachers Association, this educational reform is not lacking in supporters. The increasing number of states allowing computer science courses to be counted as credits toward graduation represents a trend that will be adopted along with the CCS curriculum reforms.

Learning to code is being promoted on the assumption that the culture, indeed, the entire world, is on the cusp of a major change that will be even more profound than the introduction of print and later literacy. In the emerging era that will be dominated by digital technologies, the ability to write computer code will lead to employment opportunities with high salaries. It is noteworthy that this effort to make learning coding skills as common as learning to think and speak is to start with students in the earliest grades.

As educational reformers have never seen a trend they did not want to embrace, the question is: Why is the effort to educate all students to write computer code an example of double bind thinking? It would seem that learning a skill that opens the door to future employment would be an unqualified gain—especially when the digital revolution is making redundant an increasing number of workers. Because teachers' mindsets, and thus interpretative frameworks, are based on the same deep cultural assumptions taken for granted by computer scientists whose interests align closely with those of the corporations that envision billions in profits from the digital revolution, it is unlikely that teachers will recognize the conceptual and moral issues that should lead to questioning the wisdom of further immersing students in the culture of computer scientists.

It has become normal for teachers, especially in the upper grades, to accept print-based abstract thinking as the most reliable source of knowledge. Their years of graduate study have conditioned them to ignore that many of the cultural patterns reinforced in face-to-face and environmental relationships are tacitly experienced and thus are largely beyond explicit awareness. Reducing the complexity and emergent nature of these culturally influenced relations (or what can be called "contexts") to what can be represented in a printed account, or to a bit of information that can be digitized, leads to a further loss of collective memory of how to live mutually supportive and community-centered lives.

Since there is a difference between how people bond in face-to-face communities and what they read, there is another issue that needs to be considered. Namely, how many teachers will be able to explain to students, even at the high school level, why abstract knowledge creates a sense of being separate from the world, and thus of being an outside observer? Does it have something to do with how sight limits awareness of the multiple forms of information communicated through the ecology of relationships present in local contexts? And how many teachers can explain the profound importance of face-to-face relationships that involve all the senses, memory, and empathy? Should students learning to write computer code understand these issues? And if students do not understand these differences, as well as other deep cultural/linguistic patterns, will learning to write computer code lead them down the same path taken by computer scientists who appear to be unaware of the ecologically and community-sustaining patterns that the digital revolution is now undermining?

Beginning in the earliest and most impressionable years in the student's education, learning to code will reproduce the same nonrelational and thus nonecological patterns of thinking that have dominated the high-status forms of knowledge in the West. Western philosophers argued about what constituted the nature and sources of knowledge, but these arguments lacked an awareness of the varied forms of knowledge created and carried forward within the oral traditions of the world's cultures. This history of equating abstract thinking with high-status knowledge, as well as with the technologies that foster abstract thinking, has put us in a double bind with the ecological systems upon which humans depend.

For example, these rationally based abstract ideas, which were not ethnographically informed, include John Locke's justification of the idea of private property, Adam Smith's theory of free markets and the "invisible hand" that ensures the progressive nature of competition, and Ayn Rand's self-interest–driven individual who celebrates the virtue of selfishness (to use the title of one of her books). These abstract thinkers, along with countless others in the West, reproduce the cultural tradition of ignoring the different cultural and natural ecologies that make this a relational and codependent world.

The ways in which courses are taught, how academic disciplines are organized to foster depth of inquiry in a narrowly focused area, as well as careers and work that are centered on a specific task, are also expressions of this cultural pattern of ignoring the relational nature of existence. Some universities are now beginning to recognize the need to adopt a more interdisciplinary perspective, which is a move away from the tradition of narrow specialization that has led to many benefits and at the same time has ensured that the problematic ecological patterns are ignored. For example, how many professors and classroom teachers are aware of how the conduit view of language (the sender/receiver model) that sustains the myth of objective knowledge makes it difficult to recognize that most words are metaphors that encode the misconceptions and silences of earlier eras?

But the narrow focus continues with promoting technologies, governmental policies, and educational reforms such as the CCS. One of the characteristics of double bind thinking is that attention is given to a single goal or policy change, ignoring the other interdependent patterns within the larger ecologies. The interconnected issues that students need to understand before learning to write computer code include the following:

(1) Because young students lack an interdisciplinary knowledge of how the world works, they need to understand the connections between language, ideologies, and the various expressions of colonization that are part of a market economy. From being unaware of the deep conceptual differences within their own culture—and largely unaware of their own taken-for-granted cultural patterns, as well as differences in and between other cultures—students learning to code will simply reproduce the same tunnel vision of computer scientists who assume that data is the basis of decision making in all cultures. Computer scientists take this narrow approach to representing what constitutes knowledge because their

previous education precluded considering why transforming the moral narratives of different cultures into a digital code cannot be achieved without gross misrepresentations. The computer scientists who are to become the models for young students to emulate also ignore moral issues within their own culture, as we can see in how they have created a culture of near total surveillance.

Their hubris and embrace of the myth of progress prevents them from asking if their coding skills have really improved people's lives in terms of making them safer from the new range of computer crime, from the hacking of the nation's infrastructure and the loss of personal information that leads to stolen identities to having their personal lives reduced to data that is being sold by data brokers to corporations and government agencies.

(2) Making decisions about what should be digitized may seem simple. As we know from observing the culture of computer scientists, personal interests in solving difficult problems, developing new computer technologies that achieve greater efficiency and predictable ways of performing a task, working with a team to make a breakthrough in achieving a better understanding, and advancing one's reputation and personal wealth all come into play. Computer scientists come into the field with a wide background of personal experience, and have mentors to guide them.

But what about the limited experiences of the grade school or high school students who are learning to write code? One of the silences in the culture of computer scientists is their lack of awareness of the cultural traditions that need to be intergenerationally renewed. Indeed, if the question is asked about which cultural values, characteristics of the cultural commons, traditions of civil liberties, and patterns of mutual support essential to living less environmentally destructive lives, few computer scientists would be able to answer it. Except for the computer scientists working with environmental scientists to collect more accurate data on changes occurring in natural systems, the majority of their colleagues have aligned their careers in ways that promote a further expansion of the consumer and surveillance culture. Computer scientists have also aligned themselves with the industrial/military complex that now works to defend the nation against the resistance now taking place in non-Western cultures to colonization by Western economic and ideological interests. The deeper issue is that the programmer's taken-for-granted cultural assumptions are always present—but hidden by the way in which print and visual models appearing on the screen hide the mindset of the people who write the programs.

It is essential that the next generation of computer programmers be aware that reading what is on the computer screen is a matter of a mind (or group of minds) meeting the mind of the student. How many classroom teachers or mentors from the culture of computer scientists possess the conceptual background necessary for explaining how taken-for-granted cultural assumptions come into play in everything appearing on the screen? How many teachers are able to recognize the assumptions of the programmer if the teacher takes for granted the same assumptions?

The tragedy is that few teachers can identify the cultural assumptions that are leading us to overshoot the sustaining capacity of natural systems. So far, I have been unable to engage teacher education faculty in an in-depth study of how the language of the curriculum, including software programs, is based on root metaphors formed in the distant past. As these root metaphors influence the choice of analogs that frame the meaning of words, there is always a problem of relying upon past ways of thinking for the solution to today's issues. The problem is made even greater by the fact that the past ways of thinking are partly responsible for the industrial/consumer-dependent lifestyle that is putting billions of tons of carbon dioxide into the atmosphere.

When the language/cultural reproduction processes are ignored, as in the current situation, students are likely to follow the conceptual patterns taken for granted by most computer scientists who ignore that progress is a myth that ignores the ecologically sustainable intergenerational forms of knowledge that cannot be digitized, and that are essential to community self-reliance.

(3) At what point are students to be introduced to the current efforts of corporations and computer scientists to computerize the workplace? How will young students respond to learning the jobs that are expected to be computerized and thus disappear within the next two decades? Will they first need to understand what the community-centered alternatives are before they address a question that seems to have escaped the attention of computer scientists working with corporations to eliminate the need for workers who must meet their most basic economic needs of food and shelter? How many teachers who are introducing students to writing computer code will be able to explain why the world's diversity of cultural commons will become increasingly important as the global economy begins to break down as a result of the further degradation of natural systems that force greater reliance upon local knowledge and patterns of mutual support?

(4) Another example of ideologically driven thinking within the culture of computer scientists that may be adopted by many students is the belief held by leading computer scientists who have morphed into futurist thinkers and now rely upon Darwin's theory of evolution to justify the cultural changes being introduced by the digital revolution. This group includes, among others, Hans Moravec, Ray Kurzweil, Peter Diamandis, and Gregory Stock. They have merged Western culture's myth of linear progress with their misinterpretation of Darwin's theory that enables them to explain how the better adapted cultural memes (computer codes rather than culturally influenced ideas and values) survive. The patterns that are excluded from their social Darwinist vision of the future, which Hans Moravec and Kurzweil also argue are the emergence of the postbiological

phase of evolution, include the diversity of the world's cultural ways of knowing. According to their reading of the data, we are on the cusp of the era of the singularity, where computers will displace humans in the process of evolution. As computer scientists represent themselves as Nature's oracle, there will be no place for cultural narratives, or for differences in cosmologies that are the basis of wisdom traditions that guide human and nonhuman relationships. Nor will there be the cultural patterns of mutual support that have not yet been monetized. The diverse cultural ecologies, which these futurist thinkers view as necessary casualties of Nature's process of natural selection, align perfectly with the ideologies of libertarianism and corporate capitalism. The implication of the evolutionary process for teachers is that there will be no need for students learning to become computer programmers to possess a knowledge that cannot be digitized of the best literary, artistic, and social justice achievements of the past. In effect, this new generation of computer programmers will be unaware of which aspects of the cultural heritage need to be renewed, and which need to be abandoned. The new generation of computer programmers, whose skills will be needed only until super computers begin to write their own programs, do not need to possess a memory of the best that previous generations of humans were able to achieve, or of the worst—the domination of others, greed, and destruction of the environment.

As self-appointed oracles of Nature's plan for replacing humans with machines, computer scientists are not likely to be concerned about the long-term fate of students seeking to acquire the skills that will help corporations increase their profits by replacing workers with robots and turning everyone into sources of data that can be used to further expand the economy. Nor are the promoters of learning to code in the classroom likely to have read the leading thinkers in their field. Thus, while learning to write computer programs cannot be separated from the larger conceptual framework that few students will be able to articulate, and thus to examine critically, they will, like others, use their skills to advance the globalization of the digital culture. This project represents a huge human experiment, as there has not been a period in human history when different cultural traditions of intergenerational knowledge and values systems have been replaced with data-based decision making. And promoting this experiment requires blindly embracing the myth of progress and existing in a hyper-mental state of hubris.

(5) Given how much of the school day will be required to cover the CCS fragments of information and skills upon which students (as well as the teacher's performance) will be tested, and given that the teacher's curricular decisions must also be covered, the question of adding the learning of computer coding skills becomes even more critical. Citizens are beginning to express their concerns about how computers are changing consciousness in ways that undermine long-term memory, and about how the growing addiction to social networks and video games is undermining the learning of the most basic face-to-face social skills.

There is a growing concern about the many political and economic issues related to the computerization of the workplace and the loss of privacy that is undermining our civil liberties.

Given the ways in which computer scientists are bringing about fundamental changes in the world's culture that have not been voted upon by local citizens who are most affected by the changes, it would seem that more Edward Snowdens would emerge from the ranks of computer scientists. The fundamental changes include creating a global network of surveillance technologies (always justified on the grounds of achieving progress in apprehending criminals and terrorists); replacing workers with computer-driven machines, and outsourcing work to regions where there are few environmental and labor protections; undermining the renewal of intergenerational knowledge, skills, and patterns of mutual support; and colonizing non-Western cultures by integrating them into a Western-dominated global economy that further serves the small percentage composed of the super rich.

Will there be enough time in the school day to engage students (especially in the upper grades) in a discussion of how to recover the traditions of local democracy and social justice traditions that are being undermined by computer scientists who were not educated to consider the unintended social consequences? Without these discussions, it is likely that students who have already been socialized to the mindset taken for granted by those promoting the digital culture will be unable to recognize what needs to be questioned. The raising of questions is the first step to resisting the further changes envisaged by computer scientists. They are already researching how to create brain implants that will improve (and control) memory, and enable the brain to be wired directly into the cloud.

Will the already crowded CCS curriculum, plus the teachers' curricular initiatives and their need to introduce students to writing computer code, leave time to learn about the changes that computer scientists are introducing that lead further down the slippery slope to a police state? Perhaps an even more basic question is whether classroom teachers are even aware of the characteristics of a police state. This is a question that brings us back to asking about the failure of universities, and specifically teacher education programs, to engage students in an in-depth study of the cultural transforming nature of technologies—including digital technologies. As the deepening ecological crisis will continue to impact more people's lives, learning about the cultural mediating nature of technologies needs to take account of what contributes to a more sustainable existence.

As I point out in *The False Promises of the Digital Revolution* (2014), universities have promoted the computer sciences in ways that have led to many important and now indispensable developments. However, these extraordinary technological achievements have not been matched by an awareness of how the new advances impact the cultural patterns of people's everyday lives. Indeed, if the word *culture* were to come up in a discussion among computer scientists (or even among most classroom teachers), it is unlikely that they would recognize the complexity and hidden depths of what this word encompasses—such as tacit forms of knowledge, the role that the culture's metaphorical language (including root metaphors) play in reproducing earlier patterns of thinking formed before there was an awareness of environmental limits, the wisdom and prejudices carried forward as part of the group's cultural commons, and the differences between cultural ways of knowing—including intergenerationally accumulated knowledge of the local bioregion. For most computer scientists, the word is simply another abstraction. Like most other abstractions such as *progress* and *individualism,* it suggests that its user is a knowledgeable thinker who does not need to bore others by mentioning the aspects of culture that cannot be digitally encoded.

Privacy is willingly given up by many students who want to share as much personal information as possible on social web sites where their network friends congregate. Later, they may become aware of the dangers and inconveniences of the loss of privacy, including the way that the prices for online purchases are adjusted by corporations that take account of known data about people's economic circumstances, and the ways police are now using data to anticipate where criminal behavior will occur; the next challenge will be to monitor people's internal thoughts. The Brave New World to which learning to code leads (in spite of its usefulness for obtaining a job) now seems to be beyond what most people question.

In 2000 Bill Joy, the cofounder of Sun Microsystems, wrote an article that appeared in *Wired* magazine titled "Why the Future Doesn't Need Us." In it he questioned the myth that only constructive gains would emerge from the headlong pursuit of scientific and technological progress. This was a challenge to the orthodoxy of his day, one which continues to be ignored, especially by educators who possess only an instrumental knowledge of how to use digital technologies. As he put it:

> The experience of the atomic scientists clearly showed the need to take personal responsibility, the danger that things will move too fast, and the way in which a process can take on a life of its own. We can, as they did, create insurmountable problems in almost no time flat. We must do more thinking up front if we are not to be similarly surprised and shocked by the consequences of our inventions....We are being propelled into a new century with no plan, no control, and no brakes....The only realistic alternative I see is relinquishment: to limit development of the technologies that are too dangerous, by limiting our pursuit of different kinds of knowledge. (Joy, 2000)

At that time, there was little awareness of how the new technologies could lead to cyber attacks on the country's energy, water, and transportation systems—and even on the banking systems. A successful attack on any part of the society's infrastructure could lead to social chaos. Unfortunately, the continued embrace of the myth of endless scientific and technological progress meant that Joy's warning has been largely ignored by computer scientists and the general public. The dangers of cyber warfare will not be of concern to young students just learning to write computer programs—especially since cyber warfare is the basis of the billion dollar video industry to which many youth are already addicted. Their thinking will mirror that of the larger society about the personal conveniences and efficiencies gained by learning to become self-directing in using digital technologies. They are also vulnerable to being caught up in the seemingly magical way the Internet can transport them to virtual realities located in different times and spaces.

Nor are they likely to be aware of the dangers connected to learning to code other people's assumptions and interpretive frameworks that they mistakenly assume to be their own original ideas and values. The important point about the long-held Western myth that equates change with a linear form of progress, as well as the transition to the time when computers surpass human intelligence, is that there is now no need to be concerned about conserving traditions, especially as both the corporations and computer scientists are racing to eliminate the need for workers.

Representing technological progress, as well as the increases in corporate profits, as governed by scientifically agreed upon natural forces that have brought all species to their present stage of evolutionary change enables futurist-oriented computer scientists to justify their silence about what needs to be conserved. If we are already entering the postbiological era dictated by the forces of evolution that have operated for billions of years, then there is no reason to be concerned about the loss of wisdom traditions that previous generations relied upon as moral guides to human-with-human and human-with-nature relationships. The current explanation of what survives in terms of cultural patterns of thinking, valuing, and behaviors is what represents the "better adapted." This awkward phrase fails to make the point of earlier social Darwinist thinkers who used the phrase "survival of the fittest"—which makes the point more clearly.

There is an irony that should be obvious to most thoughtful people. If the thinking of the computer futurists accurately represents Nature's plan—as they claim—then the classroom teachers and school administrators who are supporting learning to write computer code as a further expansion of the digital patterns of thinking are, in effect, promoting their own extinction. I suspect that it will not be the nonideological forces of natural selection that will lead to making teachers and administrators a disposable relic of the digital culture. What few realized when they gave their support to the CCS curriculum reforms, where both the pedagogy and content will be assessed in terms of a computer-derived test score, is that the long-range goal of the corporations and politicians promoting the CCS reforms is to replace teachers and administrators entirely with computer software programs. Computer programs can be tailored to reflect more accurately what various market liberal and religious extremist groups want their children to learn. For many parents who want a classical or religious-based education for their children, replacing teachers who bring a wide range of ideologically driven interpretations to the curriculum would represent an important reform.

Corporations promoting the digital revolution in education understand that the cost of production can be radically reduced by replacing humans, with their demands for health and retirement benefits and worker safety concerns, with computer-driven machines. This awareness is largely shared by segments of the public who think of education as like a consumer relationship where the primary concern is to pay the lowest price—even if the product is made in China. Replacing teachers with computer programs is simply another expression of the Wal-Mart mindset.

Administrators of major universities have also recognized that online courses and now online degrees are more profitable because the range of students that can be reached is now global. A further benefit of online degrees is that their uptake might lead to smaller enrolments in some on-campus disciplines, which allows administrators to reduce their overhead expenses by eliminating departments. The number of faculty in disciplines that lead to lower paying careers can also be reduced without undermining the success of future fundraising campaigns. Substituting computer courses for faculty, who often represent themselves as defenders of higher learning against the barbarians who happen to be the taxpayers, will appear to be social justice to many members of the public.

It will not be the new evolutionary epoch of singularity that will seal the fate of classroom teachers and school administrators. It will be a combination of corporations seeking new markets, computer scientists who are narrowly focused on keeping their jobs by coming up with new technological solutions such as the ability to store a lifetime of data on the student's educational history, and the reactionary local and national political forces that are in denial that there is an ecological crisis that by the end of the century will sweep their grandchildren away.

The market liberal forces promoting the digital revolution in education will also begin to make faculty in colleges of education redundant. Ironically, education faculty sowed the seeds that will lead to their own disappearance by ignoring the deeper cultural issues that should have included challenging the myth that technologies are culturally neutral, and that they lead only to progress. The fate of classroom teachers and even their professors will likely be of little concern as the deepening ecological crisis will focus the attention of the world's nearly 7.5 billion people on how to survive the social chaos that lies ahead.

If this seems excessively alarmist, consider what scientists are reporting about the billions of people who will shortly be facing a shortage of potable water, the rapid decline in the world's fisheries due to changes in the chemistry of the world's oceans caused by human behavior, the increasing frequency of extreme weather patterns, and the loss of habitats leading to the decline in the insect pollinators that our food supply depends upon. What the scientists are not mentioning is the social chaos that will result from the loss of intergenerational knowledge of how to live less consumer-dependent lives because the youth's reliance upon digital technologies has created the pathology of cultural amnesia. Being connected to the Internet is a mind- and thus culture-altering experience, especially when it is accompanied by the Western myth about endless consumerism, progress, and individual autonomy.

The Enlightenment vision of the West will seem all the more tragic given the suffering that will follow the abandonment of the wisdom traditions of many oral cultures that remained centered on community traditions of self-sufficiency and learning from the natural ecologies of their bioregion. Ray Kurzweil, acting as Nature's chief oracle, urges his readers not to lose faith in the Enlightenment vision of progress. It will be carried forward more efficiently by the further evolution of super-intelligent computers that will surpass all previous forms of human wisdom. Peter Diamandis and Steven Kotler, who also ignore the ecological crisis, hold out the promise that the digital revolution will lead to (to use the title of their 2012 book) *Abundance;* the subtitle promises that *The Future Is Better Than You Think.* That subtitle should be put in digitally controlled lights over the doorways where the hungry and unemployed, including classroom teachers and professors, will gather for meals and warmth.

Glossary

Definition of Terms

The dominant characteristic of ecological patterns of thinking involves taking account of the relationships and interdependencies within and between cultural and natural ecological systems. This way of thinking, within some fields, is called "systems thinking." As the word system *is too often associated with mechanistic processes and with design processes, I prefer to use the world* ecology. *What follows are definitions that should be understood as clarifying the nature and sources of misconceptions, and the cultural processes that influence patterns of thinking that, in turn, affect other cultural and natural processes. In short, ecological thinking does not focus on what are assumed to be independent entities, events, ideas, and individuals—whose "independent" status is the result of abstract thinking and technologies that are unable to represent the relational world of interactive patterns.*

Abstract thinking: Relying upon ideas derived mostly from a printed source that does not take account of the emergent nature of the ecology of direct experience, including local contexts. Abstract thinking includes many of the characteristics of print in that is provides only a surface knowledge, and it lends itself to making generalizations and statements that presume to have universal validity. It also hides the culturally influenced assumptions and silences of the person making the abstract statements or theory. Abstract thinking hides the relational and interdependent world in which we live.

Analog: As Nietzsche pointed out in the 1880s, the initial step in understanding whatever is new is to think of what it is like, or in terms of what is similar. The comparison may be derived from widely held ideas or from experience. The analog that frames how something new is to be understood often involves imposing on what is potentially unique in the old patterns of thinking. Recent examples: thinking of a computer as like the human brain, cultural patterns as like genes, data as like the objective findings of scientists, and the environment as like a natural resource.

Autonomy: The idea of autonomy, whether it is used in thinking about an ideal to be attained by the individual, about an event, or about a cultural product such as data, is an abstraction and often the result of abstract theory widely accepted by those in the grip of the liberal/industrial paradigm of earlier centuries. When it is recognized that everything, from the molecular to the cultural practices exercised in everyday life, involves relationships and interdependencies, the idea of autonomy can be recognized as an archaic metaphor. In spite of being a carryover from a period in Western thinking dominated by the abstractions of philosophers and academics who did not understand the relational and emergent nature of all ecological systems, it continues to be an abstract concept that serves as the cornerstone of current ideologies ranging from social justice to market liberalism and libertarianism—none of which take into consideration the nature and cultural roots of the ecological crisis.

Biosemiotics: A new field of study that is largely focused on the patterns of communication (semiotic processes) within biological systems. That is, it recognizes that different organisms possess different semiotic systems for communicating and responding to what is being communicated in the relationships that make up their living ecologies. Given Gregory Bateson's insight that the basic unit of information is in the "difference which makes a difference," which is a characteristic of all relationships within both natural and cultural ecologies, this new field of study should be termed ***ecosemiotics***. Both metaphors are part of an emergent vocabulary for understanding the nature of all life as ecologies, and thus the interdependency within and between ecological systems.

Conduit view of language: This phrase was introduced by Michael Reddy when he explained a widely held misconception about the role of language in human communication. The metaphor of a conduit, he explained, exemplifies how many people view the role of language as a sender/receiver process of communication. As in a conduit, ideas, data, and information are passed to others, with words passing the meaning along. The conduit view of language is essential to maintaining the idea of objective knowledge, ideas, and information because this sender/receiver view of language hides that words have a history, and that they are metaphors that carry forward the meaning framed by the analogs settled upon in the distant past.

Conservative: When used to refer to cultural traditions, it means carrying forward past ways of thinking, cultural patterns, and achievements, as well as social injustices and prejudices. When used by environmentalists, it means conserving habitats and species. When used to refer to an ideology, it means being mindful about which aspects of the past should be intergenerationally renewed, such as separation of church and state, civil liberties, local democracy, social justice achievements, and past achievements in the arts and sciences, as well as cultural practices that strengthen community. The current Orwellian use of *conservatism* means supporting market liberal and libertarian values and policies where everything is understood as having only an economic value.

Cultural commons: The phrase "cultural commons" refers to the largely nonmonetized, intergenerational knowledge, skills, and mentoring relationships that enable people to participate in mutually supportive communities. Every culture has its cultural commons traditions that are dependent upon face-to-face relationships, that provide the young with opportunities to discover talents and develop skills across a wide range of activities, from preparing and sharing food and learning the medicinal knowledge acquired over generations to building on the traditions in the arts, crafts, and uses of technology that reflect a knowledge of local ecosystems. They include games, ceremonies, narratives, traditions of social justice (and for some cultural commons, traditions based on prejudices and patterns of exclusion), and language itself. The cultural commons are as diverse as the languages still spoken in the world, and they exist in urban and rural areas, as well as in every community and family. In short, they represent activities and relations that contribute to communities that are mutually supportive and that have a smaller carbon footprint. In short, they represent community sites that represent alternatives to a consumer-dependent and environmentally destructive lifestyle. They also represent alternative spaces where wealth is measured in returning to the community the talents, skills, and mentoring relations that enrich and support others, rather than in terms of material riches. Given this way of understanding wealth, poverty takes on a new meaning: namely, as not possessing skills and talents that make a contribution to the well-being of the community.

Difference which makes a difference: This phrase was used by Gregory Bateson to explain how both natural and cultural ecologies involve relationships—with the differences occurring within these relationships being the sources of information (or signs) that lead to responses that become, in turn, the difference which makes a difference. The silence in a conversation becomes a difference which makes a difference to the other participants, which leads them to respond, which then leads to differences that continue to change the relationship. What this phase helps us understand is that all life processes, from the molecular to global weather systems, are dynamic semiotic systems that influence and respond to the patterns that connect. In short, differences are the most basic and primary sources of information circulating through all systems. Put another way, everything exists in an ecology of relationships—that is, nothing is totally independent and autonomous. This holds for facts, objective data, universal truths, ideas, values, and autonomous individuals.

Double bind: This phrase was introduced by Gregory Bateson to describe how therapists, thinking they are offering helpful advice, often contribute to the problems they are attempting to solve. More generally, double bind thinking occurs when what is assumed to be a progressive development or policy solution actually further exacerbates the problem. Current examples of double bind thinking include promoting economic growth that further deepens the ecological crisis, and developing surveillance technologies that supposedly safeguard Americans from terrorists but undermine American civil liberties. The double bind thinking that underlies the justification of the Common Core Standards is that knowledge and skills that are to contribute to creating a more reliable workforce do not take into account that corporations are automating the workplace—which will eliminate the need for an estimated 47% of workers over the next two decades.

Ecologies—both natural and cultural: Until recently, ecologies were understood by scientists as natural systems. More recently, ecologies are understood as including all aspects of culture, such as the ecology of language, identities, mutual support systems, families, market systems, wars, terrorism, and so forth. What is distinctive about ecologies, both natural and cultural, is that the historical, present, and future prospects of the ecological system are taken into consideration. The focus is on past, present, and future relationships that influence the life of the ecological system. Not all ecologies lead to the renewal and sustainability of the system.

Ecological intelligence: This phrase refers to the patterns of thinking/awareness/behavior that involve reliance upon the five senses, memory, recognition of taken-for-granted cultural patterns, and the differences communicated by other participants in the local natural and cultural ecologies. It is the opposite of the myth where the individual thinks and acts upon an external world without considering the complex patterns of communication. The person playing a game of chess, playing tennis, swimming, engaged in a dialogue with others, cooking a curry, giving close attention to traffic patterns, and so forth is exercising a degree of ecological intelligence. This is the opposite of the person engaged in a monologue with others, imposing ideas and actions on the environment or others that are based on preconceived plans and abstract ideas. There are different degrees of exercising ecological intelligence. The first is individually centered, where attention is given to what needs to be taken into account in order to achieve one's own goals. The second level involves awareness of the differences which make a difference in social relationships—with particular attention being given to the patterns of unjust relationships and to strategies that lead to reforms. The third level of ecological intelligence involves awareness of how different elements in the cultural and natural ecologies contribute to or undermine the long-term prospects of an ecologically sustainable future. Revitalizing the cultural commons, using locally adapted technologies, reducing dependence upon consumerism, the practice of local democracy, participating in the slow-food movement, being aware of how language carries forward the misconceptions that underlie the Industrial Revolution and a consumer-dependent lifestyle, and so forth, are all expressions of sustainable ecological intelligence.

Ecojustice: The five guiding principles that serve as the starting point for understanding ecojustice in the 21st century include: (1) ensuring that marginalized groups are not subjected to the health risks of living close to industrial sites; (2) pursuing policies that do not exploit the resources of non-Western cultures—and lead to colonizing them to adopt a consumer-dependent lifestyle; (3) revitalizing what remains of the diversity of the world's cultural commons that exist in every community, and that represent the intergenerational alternatives to dependence upon consumerism that degrades the environment and leads to various forms of impoverishment; (4) living less consumer-dependent lifestyles that impact natural systems in ways that limit the prospects of future generations to live full and mutually supportive lives; (5) and recognizing what Vandana Shiva refers to as "earth democracy": that is, the right of species to reproduce themselves and to participate in the earth's ecosystems without being reduced to an exploitable resource.

Enclosure: The history of this word in the West comes from the mid-19th-century enclosure laws of the British parliament, which transformed the tradition of land (the natural environment) being freely available to everyone to use into a system of privately owned lands that could be used for specific agricultural practices needed to power the Industrial Revolution. This historical meaning is retained today, as the enclosure of the cultural commons involves transforming what is freely shared into what is privately owned, and into the market system that forces people to be dependent upon a money system that magnifies inequalities and dependencies. Transformation, exclusion, silences, privatization, and monetization are the chief characteristics of the enclosure of the cultural commons.

Freedom: Existentially, everyone exercises a limited degree of freedom in how they interpret the responses of the Other, in whether they think reflectively or accept the taken-for-granted world of others, and in making a wide range of decisions within the conceptual and moral constraints of their cultures. The fact is that consciousness, and thus the ability to think and communicate, is influenced by the language community into which the individual is born. This means that the individual can never entirely escape the influence of her/his culture—and thus should never be considered free in the sense the word *freedom* is often used by educational reformers and politicians. Life always involves relationships, and the choices within these relationships lead to biographically distinct personal histories, but never to freedom in the abstract sense of the word. Franklin D. Roosevelt's Four Freedoms speech to Congress in 1941 represents yet another way in which the metaphor can be used to address social justice issues.

Globalization: The digital revolution, the market liberal ideology that guides corporate policies, and the idea of global military preparedness are the major forces behind globalization. Globalization also has roots in the early missionary and also current idea of American exceptionalism that leads Americans to assume they have the right to impose their cultural values on the rest of the world. Computer scientists view globalization as leading to all cultures becoming dependent upon the Internet as their primary source of knowledge, while educational reformers continue to promote a critical inquiry mode of consciousness that supposedly leads to individual autonomy. Not considered are the forms of knowledge and traditions of different cultures that enable them to live less consumer-dependent lifestyles—and in many instances, less environmentally destructive lifestyles. Globalization is often justified on the grounds that it improves people's economic circumstances (which it does, in some situations), but less attention is given to the impact of globalization on the world's diversity of local cultural commons that enable people to live less consumer-dependent lives.

Ideology: As Clifford Geertz points out, ideologies are interpretative frameworks that guide a wide range of decisions on how social life should be organized, reformed, and used as a basis for colonizing other cultures. They have a history that can be traced back to the deep assumptions of a culture, which are often the basis of thinking of philosophers and other theorists who provided the conceptual foundations for these abstract ideologies that are ecologically and ethnically uninformed. Marxism, socialism, liberalism, libertarianism, social Darwinism, and print-based religious traditionalism are examples of current ideologies. Environmental conservatism is an ideology based on the scientific studies of the destruction of habitats and the loss of species.

Information pathways: This phrase highlights the importance of recognizing the relational nature of cultural and natural ecologies, and refers directly to the semiotic (sign systems) capacities of different life-forming and -regulating systems within these interdependent ecologies. Cultural patterns of thinking can impede awareness of the different ways in which ecological systems communicate about their states of being. These include the misconceptions encoded in the metaphorical language that represents humans as at the center of the world, and intelligence as free of past linguistic influences and awareness of environmental limits. Other cultural influences that marginalize awareness of what is being communicated through different information pathways of other organisms, including natural systems such as global warming and the acidification of the world's oceans, include the silences in the media and the educational systems, and ideologies that promote consumerism and economic growth. The information pathways are the same as the differences which make a difference within ecological systems.

Libertarianism: This is an ideology heavily influenced by the writings of Ayn Rand, but also by the abstract thinking of a number of Western philosophers who were ethnocentric thinkers. The key ideas include the notion that individual autonomy and well-being should govern all decisions, that unrestricted capitalism should be the basis of social life, and that government should not provide a safety net for those in need. Being in need is assumed by libertarians to be evidence of a fault in the individual's character. Thus, the role of government is limited to national defense and protecting contracts, and does not extend to correcting the faults of the individual's character. This ideology is closely aligned with social Darwinist thinking, as well as with market liberalism. It also ignores the ecological crisis and the imperialistic agenda of American foreign policies, and makes a virtue of individual selfishness. It has many adherents in the United States.

Market liberalism: Market liberalism is incorrectly called "conservatism"—which is more than ironic because its primary agenda is to expand markets and profits while undermining what remains of the cultural commons, the Constitution, and the democratic process. Its core ideas are based on a number of deeply held Western cultural assumptions (which can be called "root metaphors"). These include the autonomous nature of the individual, that this a human-centered world, that change (especially technological change) is an inherently progressive force, that technology is culturally neutral and essential to social development, that every aspect of life can be improved by being monetized and subjected to market forces, and that this is a mechanistic world that can be engineered in ways that lead to more profits. It is the opposite of the conservatism of environmentalists, proponents of the cultural commons that include the traditions of civil liberties and the Constitution. Like libertarianism, it often aligns itself with the pseudo-scientific theory of social Darwinism.

Mediator: This word is used to describe the role of the teacher in helping students recognize the fundamental differences between their cultural commons and consumer and monetized forms of dependency. As a mediator, the teacher is not to be engaged in indoctrinating students to adopt either position, but rather to raise questions about the students' experiences that they may not have considered. Like the mediator in a labor dispute, the teacher should be knowledgeable about both lifestyles, and thus be able to ask questions that lead students to examine the differences in terms of which has a smaller ecological impact, strengthens patterns of mutual support, leads to the discovery and development of personal talents, and contributes to a sense of community and mutually supportive relationships. As a mediator, the teacher's role in helping students examine their otherwise taken-for-granted experiences as they move between cultural commons and market-based experiences is to help them acquire the language necessary for naming and understanding relationships and patterns that previously were part of the silences imposed by the curriculum of public schools and universities. In the past, teachers who helped students recognize and articulate the taken-for-granted cultural patterns that supported patriarchal and racist traditions were exercising their responsibility as mediators in achieving a more just society. Now when teachers focus on helping students clarify the differences between their own as well as other people's cultural commons and market forms of dependencies, they are also contributing to an ecojustice future.

Metaphorical language and thinking: This phrase refers to a basic aspect of language that is overlooked partly because of the taken-for-granted misconceptions surrounding the conduit view of language. Basically, it highlights that most words are metaphors that carry forward the assumptions, misconceptions, and silences of earlier times. In many instances, the analogs that continue to frame the current meanings of words (metaphors) such as *individualism, traditions, intelligence, environment, data, progress, intelligence,* and so forth, reflect the culturally influenced thinking of earlier eras. When students are not informed that words have a history that carries forward earlier misconceptions and even important insights, their taken-for-granted patterns of thinking are largely influenced by earlier forms of cultural intelligence which may be totally inadequate for understanding the issues we face today. Teachers can address this problem by encouraging students to examine the cultural history of words, the ways in which the linguistic colonization of other cultures occurs, and the differences between print-based and largely oral-dependent cultures.

Objective knowledge and data: The metaphor of *objective* reproduces the cultural myth that it is possible to know something that is not influenced by the language of the observer and her/his taken-for-granted culturally influenced patterns of thinking. The conduit view of language, which is essential to sustaining the myth of objectivity by hiding that words have a history, helps to sustain the myth. The myth of objective knowledge is one of the most overlooked exercises of power, as the claim of objectivity delegitimizes other ways of knowing.

Root metaphors: Mythopoetic narratives and powerful evocative experiences have become the root metaphors that serve as largely taken-for-granted interpretative frameworks that explain a wide range of cultural behaviors. In the West, the root metaphors of patriarchy and a human-centered view of the world have been widely and erroneously associated with the **Book of Genesis**. Other consciousness and thus culture-shaping root metaphors include individualism, mechanism, progress, economism, evolution, and now, ecology. The latter now challenges these earlier root metaphors that underlie the modern industrial/consumer-dependent culture that is being globalized. Root metaphors may be mutually supportive, such as individualism, mechanism, progress, economism, and now, evolution (which underlies the West's approach to development). Another characteristic of root metaphors is that they exclude alternative vocabularies that challenge the interpretative framework of the dominant root metaphors. The mechanism of root metaphor makes no allowance for words referring to social justice; the vocabulary of the root metaphor of progress represents traditions as sources of backwardness, and the root metaphor of evolution represents the vocabularies of the sacred and wisdom traditions as examples of an earlier stage of evolutionary development and thus as archaic. Basically, the prevailing root metaphors influence the choice of analogs that frame the meaning of other parts of the culture's vocabulary. Relying upon the metaphors framed by the misconceptions of the past as the basis for addressing the cultural roots of the ecological crisis is yet another example of double bind thinking.

References

Bateson, G. (1972). *Steps to an ecology of mind.* New York: Ballantine.

Berry, W. 2000. Life is a Miracle. Washington, D.C.: Counterpoint Press

Bowers, C. (2011). *Perspectives on the ideas of Gregory Bateson, ecological intelligence, and educational reforms.* Eugene, OR: Eco-Justice Press.

———. (2014). *The false promises of the digital revolution: How computers are changing education, work, and international development in ways that are ecologically unsustainable.* New York: Peter Lang.

Common Core State Standards Initiative http://www.corestandards.org/wp-content/uploads/ELA_Standards.pdf

Crick, F. (1995). *The astonishing hypothesis: The scientific search for the soul.* New York: Scribner.

Diamandis, P., & Kotler, S. (2012). *Abundance: The future is better than you think.* New York: Free Press.

Frey, C., & Osborne, G. (2013). *The future of employment: How susceptible are jobs to computerization?* Available at http://www.oxfordmartin.ox.ac.uk/downloads/academic/The_Future_of_Employment.pdf

Geertz, C. (1977). *The interpretation of cultures.* New York: Basic Books.

Hoffmeyer, J. (2008). *A legacy for living systems: Gregory Bateson as precursor to biosemiotics.* Dordrecht: Springer.

Idhe, D. (1979). *Technics and praxis: A philosophy of technology.* Dordrecht: Springer.

Joy, B. (2000). "Why the future doesn't need us." *Wired 8,* 4. Available at http://archive.wired.com/wired/archive/8.04/joy.html

Kurzweil, R. (1999). *The age of spiritual machines: When computers exceed human intelligence*. New York: Viking.

———. (2005). *The singularity is near: When humans transcend biology*. New York: Viking.

———. (2012). *How to create a mind: The secret of human thought revealed.* New York: Viking.

Maybury-Lewis, D. (1992). *Millennium: Tribal wisdom and the modern world.* New York: Penguin Books.

Moore, T. (2013). *The story killers: A common-sense case against the Common Core.* Amazon's CreateSpace Independent Publishing.

Moravec, H. (1990). *Mind children: The future of robots and human intelligence.* Cambridge, MA: Harvard University Press.

Nisbett, R. (2003). *The geography of thought: How Asians and Westerners think differently…and why*. New York: Free Press.

Sagan, C. (1997). *The demon-haunted world: Science as a candle in the dark.* New York: Random House.

Shabecoff, P., & Shabecoff, A. (2008). *Poisoned profits: The toxic assault on children.* New York: Random House.

Smith, H. (1994). *The illustrated world's religions: A guide to our wisdom traditions.* New York: HarperCollins Publishers.

Spretnak, C. (2011). *Relational reality: New discoveries of interrelatedness that are transforming the world.* Topsham, ME: Green Horizon Books.

Stock, G. (1993). *Metaman: The merging of humans and machines into a global superorganism*. New York: Doubleday.

Whitehead, A. (2010). *Process and reality.* New York: Simon & Schuster. (Original work published 1929)

Wilson, E. (1998). *Consilience: The unity of knowledge*. New York: Alfred A. Knopf.

Also by C. A. Bowers

The Promise of Theory: Education and the Politics of Cultural Change

Elements of a Post-Liberal Theory of Education

The Cultural Dimensions of Educational Computing: Understanding the Non-Neutrality of Technology (with David Flinders)

Responsive Teaching: An Ecological Approach to Classroom Patterns of Language, Culture, and Thought

Education, Cultural Myths, and the Ecological Crisis: Toward Deep Changes

Critical Essays on Education, Modernity, and the Recovery of the Ecological Imperative

Educating for an Ecologically Sustainable Culture: Re-thinking Moral Education, Creativity, Intelligence, and Other Modern Orthodoxies

The Culture of Denial: Why the Environmental Movement Needs a Strategy for Reforming Universities and Public Schools

Let Them Eat Data: How Computers Affect Education, Cultural Diversity, and Prospects of Ecological Sustainability

Educating for Eco-Justice and Community

Detrás de la Apariencia: Hacia la Descolonización de la Educación

Mindful Conservatism: Rethinking the Ideological and Educational Basis of of an Ecologically Sustainable Future

Rethinking Freire: Globalization and the Environmental Crisis (co-edited with Frédérique Apffel-Marglin)

The False Promises of Constructivist Theories of Learning

Revitalizing the Commons: Cultural and Educational Sites of Resistance and Affirmation

Perspectives on the Ideas of Gregory Bateson, Ecological Intelligence, and Educational Reforms

University Reform in an Era of Global Warming

Educational Reforms for the 21st Century

The Way Forward:Educational Reforms that Focus on the Cultural Common and the Linguistic Roots of the Ecologial/Cultural Crises